Recombinant Antibodies

Recombinant Antibodies

Applications in Plant Science and Plant Pathology

KAREN HARPER AND ANGELIKA ZIEGLER

Scottish Crop Research Institute, Dundee, UK

TAYLOR & FRANCIS
ALERE FLAMMAM
Founded 1798

First published 1999 by Taylor & Francis
11 New Fetter Lane, London EC4P 4EE

Simultaneously published in the USA and Canada
by Taylor & Francis Inc,
325 Chestnut Street, 8th Floor, Philadelphia PA 19106

Taylor & Francis is an imprint of the Taylor & Francis Group

Typeset in Times by Graphicraft Limited, Hong Kong
Printed and bound in Great Britain by T.J. International, Padstow, Cornwall

Every effort has been made to ensure that the advice and information in this book is true and accurate at the time of going to press. However, neither the publisher nor the authors can accept any legal responsibility or liability for any errors or omissions that may be made. In the case of drug administration, any medical procedure or the use of technical equipment mentioned within this book, you are strongly advised to consult the manufacturer's guidelines.

British Library Cataloguing in Publication Data
A catalogue record for this book is available from the British Library

Library of Congress Cataloging in Publication Data are available

ISBN 0-7484-0763-4

Contents

Preface

The range of applications of antibodies to plant science and plant pathology has grown rapidly over the last 20 years. More recently, recombinant antibody fragments – derived from existing hybridoma cell lines or selected from phage display libraries – have appeared on the scene and are now becoming more widely used.

In the first part of this book, some background information on antibodies and their interaction with antigens is provided. The production of recombinant antibodies from hybridoma cell lines and phage display libraries is described, and examples are given of how they may be used as diagnostic tools in plant pathology. Also, modern methods for analysing binding properties of antibodies are reviewed.

The second part of this book explores the use and potential of recombinant antibodies for engineering resistance to plant pathogens and pests, and for immunomodulation of plant metabolism. Also, the potential of plants as 'factories' for the production of recombinant antibodies is explored.

We hope this book draws attention to, and highlights some of, the exciting possibilities offered by recombinant antibody technology to those working in plant science.

We would like to thank the authors for their valuable contributions.

Finally, we thank all our colleagues in the Virology Department of the Scottish Crop Research Institute for their help and understanding while this book was being compiled.

K. Harper & A. Ziegler

Contributors

O. Artsaenko
University of Maryland, Baltimore County
Chemical and Biochemical Engineering
1000 Hilltop Circle
Baltimore, MD 21250
USA

J. Bakker
Department of Nematology
PO Box 8123
6701 ES Wageningen
The Netherlands

U. Conrad
Institut für Pflanzengenetik und Kulturpflanzenforschung Gatersleben
06466 Gatersleben, Corrensstr. 3
Germany

G. De Jaeger
Department of Plant Molecular Genetics
University of Gent (VIB)
K.L. Ledeganckstraat 35
9000 Gent
Belgium

M. De Neve
Department of Plant Molecular Genetics
University of Gent (VIB)
K.L. Ledeganckstraat 35
9000 Gent
Belgium

A. Depicker
Department of Plant Molecular Genetics
University of Gent (VIB)
K.L. Ledeganckstraat 35
9000 Gent
Belgium

C. De Wilde
Department of Plant Molecular Genetics
University of Gent (VIB)
K.L. Ledeganckstraat 35
9000 Gent
Belgium

L. Fecker
Institut für Pflanzengenetik und Kulturpflanzenforschung Gatersleben
06466 Gatersleben, Corrensstr. 3
Germany

U. Fiedler
Martin-Luther-Universitat Halle-Wittenberg
Fachbereich Biochemie/Biotechnologie, Institut für Biotechnologie
06120 Halle (S.), Kurt-Mothes-Str. 3
Germany

F. Gommers
Department of Nematology
PO Box 8123
6701 ES Wageningen
The Netherlands

K. Harper
Department of Virology, Scottish Crop Research Institute
Invergowrie
Dundee DD2 5DA
UK

G. Himmler
Hygiene Institute
University of Vienna
Austria

R. Kerschbaumer
Institute of Applied Microbiology
University of Agriculture and Forestry
1190 Vienna, Nussdorfer Lände 11
Austria

R. Koenig
Institut für Biochemie und Pflanzenvirologie
Biologische Bundesanstalt Braunschweig
38104 Braunschweig, Messeweg 11–12
Germany

M.A. Mayo
Department of Virology, Scottish Crop Research Institute
Invergowrie
Dundee DD2 5DA
UK

M. Peisker
Institut für Pflanzengenetik und Kulturpflanzenforschung Gatersleben
06466 Gatersleben, Corrensstr. 3
Germany

J. Phillips
QIAGEN GmbH,
Max-Volmer-Str.4
40724 Hilden
Germany

A. Schots
Department of Nematology
PO Box 8123
6701 ES Wageningen
The Netherlands

P. Seddas-Dozolme
Laboratoire de Pathologie Vegetale, Unite de Recherche Vigne et Vin
INRA Colmar
28 Rue de Herrlisheim
68021 Colmar
France

G. Smant
Department of Nematology
PO Box 8123
6701 ES Wageningen
The Netherlands

L. Torrance
Department of Virology, Scottish Crop Research Institute
Invergowrie
Dundee DD2 5DA
UK

R.L. Toth
Department of Virology, Scottish Crop Research Institute
Invergowrie
Dundee DD2 5DA
UK

M. Van Montagu
Department of Plant Molecular Genetics
University of Gent (VIB)
K.L. Ledeganckstraat 35
9000 Gent
Belgium

M.H.V. Van Regenmortel
Laboratoire d'Immunochimie des Peptides et des Virus
IBMC
15 rue Rene Descartes
67084 Strasbourg Cedex
France

B. Walter
Laboratoire de Pathologie Vegetale, Unite de Recherche Vigne et Vin
INRA Colmar
28 Rue de Herrlisheim
68021 Colmar
France

G. Zeder-Lutz
Laboratoire d'Immunochimie des Peptides et des Virus
IBMC
15 rue Rene Descartes
67084 Strasbourg Cedex
France

A. Ziegler
Department of Virology, Scottish Crop Research Institute
Invergowrie
Dundee DD2 5DA
UK

PART ONE

Recombinant Antibodies as Diagnostic Tools in Plant Pathology

1

Introduction: Antigens, Antibodies and Plantibodies

PASCALE SEDDAS-DOZOLME, BERNARD WALTER AND
MARC H.V. VAN REGENMORTEL

1.1 Introduction

In order to understand how antibodies can be applied in plant science and plant pathology, it is necessary to have some insight into our current knowledge of immunochemistry. The purpose of this chapter is to provide plant scientists and plant pathologists with some basic information on antigens and antibodies and on the process of antigen–antibody interaction. Plantibodies are recombinant antibodies expressed in plants which possess the same antigen binding capacity as the parent monoclonal antibody.

1.2 Antigens

The term antigen refers to any entity, whether a cell, a macromolecular assembly or a single molecule, which can elicit an immune response in a competent host and be recognized by the products of that immune response. Antigenic reactivity or antigenicity refers to the ability to be recognized by antibodies, whereas immunogenicity refers to the ability to elicit an immune response, a property which depends on the potentialities of the animal being immunized. Since most antigens of biological interest are proteins, the discussion of antigens in this chapter will be limited to proteins.

The antigenic reactivity of a protein is located in discrete regions of the molecule known as antigenic determinants, or epitopes, which react specifically with antibodies to the protein. Each epitope is delimited by an antibody footprint of about 800 $Å^2$ and consists of an array of atoms juxtaposed on the protein's three-dimensional surface. The cluster of atoms forming an epitope can be identified only through its capacity to bind to a complementary antibody combining site or paratope. Similarly, the paratope can be defined only by referring to its complementary epitope. Epitopes and paratopes are thus relational entities definable only by their mutual complementarity and not by intrinsic features of each partner existing independently of this relationship (Van Regenmortel, 1994).

Initially, epitopes were classified as sequential or conformational, depending on whether or not they could be mimicked in active form by linear peptides. At present, epitopes are usually classified as either continuous or discontinuous, depending on whether the amino

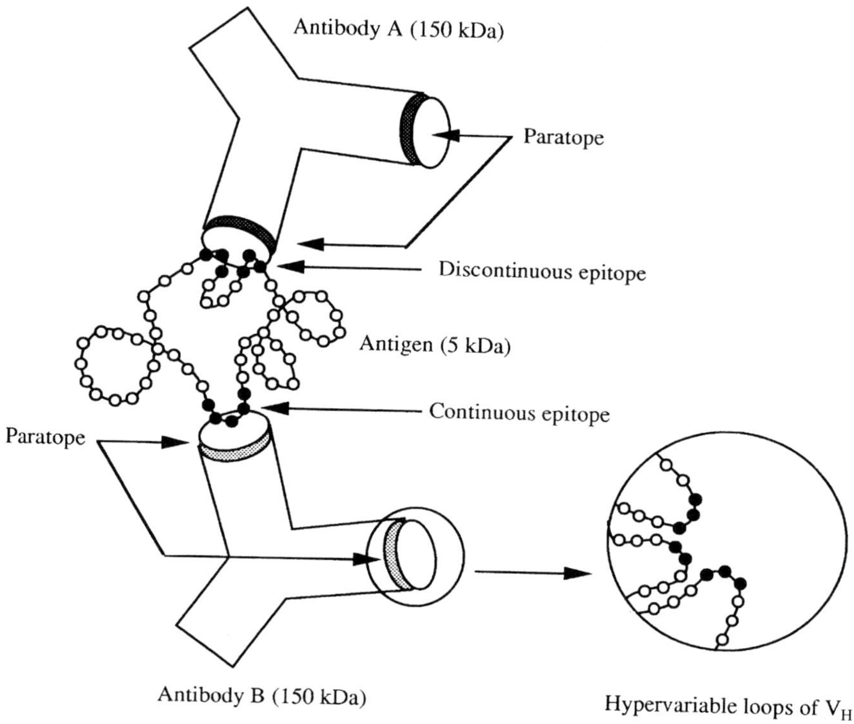

Figure 1.1 Interaction of two antibodies with a continuous and a discontinuous epitope of an antigen. Interacting amino acid residues are indicated in black.

acid residues in the epitope are contiguous in the polypeptide chain or not (Figure 1.1). Continuous epitopes correspond to short peptide fragments of a protein that can be shown to bind to antibodies raised against the intact protein. Discontinuous epitopes are made up of residues that are not contiguous in the sequence but are brought close together by the folding of the peptide chain. Although this classification is widely used, there is, in fact, no clear-cut borderline between continuous and discontinuous epitopes. In general, discontinuous epitopes contain several segments of a few contiguous residues and some of these may be able on their own to bind to the paratope. In such a case, the segments may be given the status of continuous epitopes in spite of being part of a larger discontinuous epitope. Conversely, so-called continuous epitopes usually contain a number of residues that are not implicated in the binding reaction and can be replaced by any of the other 19 possible amino acids without impairing the antigenic reactivity of the peptide. This makes the continuous epitope, functionally speaking, discontinuous. In practice, whenever a peptide of up to about ten residues is found to bind to an antibody, it is called a continuous epitope.

Regarding the number of epitopes likely to be found on a protein molecule, it is now generally accepted that the entire accessible surface of a protein consists of a large number of overlapping epitopes (Benjamin *et al.*, 1984). In view of the relational nature of epitopes, the number of epitopes in a protein can be estimated from the total number of different monoclonal antibodies (Mabs) that can be raised against the protein. On this basis, it has been estimated that the insulin molecule possesses at least 115 epitopes.

It must be stressed that it is impossible to draw sharp boundaries between the many individual overlapping epitopes which collectively can be viewed as forming an antigenic

Table 1.1 Methods used for localizing epitopes in proteins

Method	Number of residues in epitope	Criterion for residue allocation
Crystallographic analysis of antigen–antibody complexes	15–22	Contact in epitope–paratope interface
Synthetic peptides as antigenic probes in binding studies Free peptides Peptides adsorbed to solid-phase Peptides conjugated to a carrier Peptides attached to support used for synthesis	5–8	Cross-reactive binding with antiprotein antibodies
Identification of critical residues in peptides by systematic replacement of residues	3–5	Decrease in cross-reactive binding
Peptide sequences inserted in recombinant proteins	5–8	Cross-reactive binding with antiprotein antibodies
Binding of antipeptide antibodies to protein	5–8	Induction of cross-reactive antibodies
Binding of antiprotein antibodies to chemically modified or mutated proteins	3–5	Decrease in binding compared with unmodified protein
Protection of antibody-bound residues of epitope to chemical attack	5–10	Chemical reactivity of individual residues
Competitive topographic mapping	Only relative position of epitope is defined	Not applicable

site on the surface of the protein. The same residues of a protein antigen may be involved in a number of different epitopes built up from individual atomic groups that interact in a unique manner with paratopes showing no sequence similarity (Lescar *et al.*, 1995). Conversely, the same antibody may sometimes recognize two epitopes devoid of any sequence similarity. Such findings are due to the fact that the recognition between epitope and paratope does not take place at the level of whole residues but at the level of individual atoms. It is indeed most unlikely that all or even most of the atoms of a given residue will participate in a specific interaction. When an antibody recognizes a slightly different cross-reacting epitope in a related antigen, the relational link between epitope and paratope means that it is not exactly the same paratope that is involved in the cross-reaction since at least some of the atomic interactions will be different.

The methods that have been used for identifying epitopes in proteins are listed in Table 1.1. The physico-chemical principles behind these various approaches have been described elsewhere (Van Regenmortel, 1994) and detailed experimental protocols used for epitope mapping have been published (Morris, 1996).

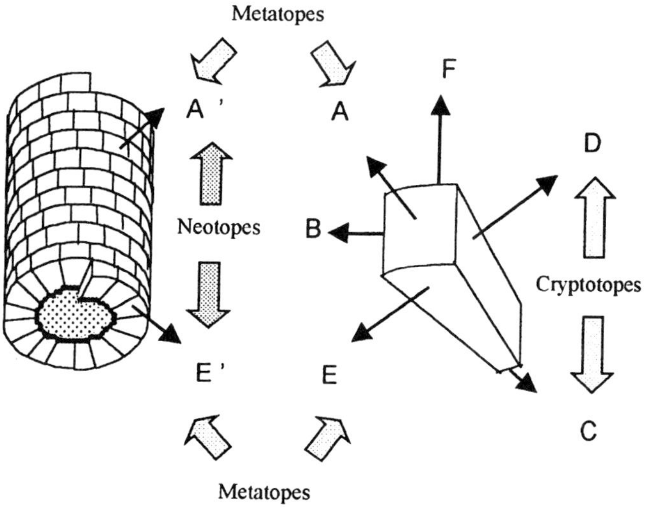

Figure 1.2 Schematic model of TMV protein subunits in monomeric form and in the virus particle, with an indication of which protein surface is recognized by various Mabs. This allocation was done on the basis of BIAcore mapping results (Saunal and Van Regenmortel, 1995).

When discussing viral epitopes, it is useful to distinguish four categories of epitopes, i.e., cryptotopes, neotopes, metatopes and neutralization epitopes. The first three categories are illustrated in Figure 1.2 in the case of the protein subunits of tobacco mosaic virus (TMV). This virus consists of 2130 identical subunits arranged as a helix around an RNA molecule. Several distinct epitopes have been identified on the protein subunit (TMVP) which comprises 158 amino acid residues (Van Regenmortel *et al.*, 1993).

Cryptotopes are hidden epitopes buried inside the assembled virus particle which become accessible to antibodies only after dissociation of the virion. A number of cryptotopes have been located in the TMVP molecule by showing that certain peptide fragments of the protein were recognized by specific antibodies (Van Regenmortel *et al.*, 1993).

Neotopes are specific to the quaternary structure of the virus particle and are not present at the surface of monomeric subunits. Neotopes arise either through conformational changes in the monomers induced by intersubunit bonds or through the juxtaposition of residues from neighbouring subunits. Metatopes are epitopes that are present in both dissociated and polymerized forms of the viral protein. Some information on the location of neotopes and metatopes on TMV particles and TMVP subunits has been obtained by immunoelectron microscopy (Dore *et al.*, 1990) and biosensor technology (Saunal and Van Regenmortel, 1995). The surfaces of the TMVP subunit on which neotopes and metatopes have been identified are shown in Figure 1.2.

Neutralization epitopes correspond to epitopes of the virus that are recognized by antibody molecules able to neutralize the infectivity of the virus (Dimmock, 1993). Such epitopes can only be identified in a functional assay, the neutralization test, and their properties can only be analysed within the operational context of a neutralization assay. In the case of plant viruses, neutralization of infectivity can be assessed in an *in vitro* biological assay in which the number of local lesions induced by inoculating leaves with the virus is compared before and after mixing infectious virus with antibody. Such a local lesion assay is similar to the plaque reduction test used in animal virology and the antibody-induced reduction in the number of lesions is clearly a neutralization phenomenon.

Neutralization of a plant virus *in vivo* may occur when antibodies expressed in plants interfere with the disassembly of virions and the translation of viral RNA. It has been shown that disassembly of TMV particles can be induced by ribosomes that reach the 5′ leader sequence of the viral RNA. These ribosomes initiate translation and dislodge the coat protein subunits during their translocation on the viral RNA (Wilson, 1984; Shaw *et al.*, 1986). Some Mabs that react with the extremity of the TMV particle containing the 5′ end of the RNA are able to block the disassembly of TMV by ribosomes, apparently by sterically preventing the interaction between virus and ribosomes (Saunal and Van Regenmortel, 1995). Such antibodies, when expressed in plants, may be useful for controlling virus infection. Another potential method for inhibiting viral multiplication would be to express in plants antibodies specific for the viral movement proteins known to facilitate the passage of certain viruses from cell to cell (Hull, 1989; Deom *et al.*, 1992; Mushegian and Koonin, 1993).

1.3 Immunoglobulins, antibodies and paratopes

Antibodies belong to a family of glycoproteins known as immunoglobulins, which are found in the fraction of serum known as the gamma globulin fraction. Immunoglobulins are heterodimers consisting of four polypeptide chains linked by disulphide bridges. There are two identical heavy (H) chains of about 450–600 amino acid residues and two light (L) chains of about 220 amino acid residues. Each H and L chain consists of homologous segments of about 110 amino acid residues that form independently folded domains (Figure 1.3). H and L chains all contain a variable region and a constant region, which are encoded by separate gene segments. The amino acid sequence of the amino terminal domain of H and L chains varies in antibodies of different specificity. Within these variable (V) domains of the H and L chains, three segments exhibit hypervariability and constitute the complementarity-determining regions (CDR) of the immunoglobulin. The six CDRs which comprise about 50 amino acid residues form hypervariable loops at the ends of the two variable domains, which are labelled L1, L2 and L3 for the light chain and H1, H2 and H3 for the heavy chain (Figure 1.3).

The remaining domains in each chain are invariant and are called constant (C) regions. In the H chain of the most common type of immunoglobulin known as IgG, there are three constant domains labelled C_{H1}, C_{H2} and C_{H3}. Between the C_{H1} and C_{H2} domains, there is an additional segment known as the hinge which contains the inter-heavy chain disulphide bridges and confers a certain degree of flexibility on the molecule. The L chains contain a single constant domain called C_L. Immunoglobulins also contain several carbohydrate chains that are usually linked to H chains. Membrane forms of immunoglobulins have additional peptide segments at the C-terminal end of their H chains which allow them to be embedded in cell membranes (Nezlin, 1994).

Immunoglobulins can be cleaved into various fragments by the action of proteases. Papain cleaves the heavy chain of IgG at the N-terminal side of the disulphide bridges that keep the H chains together, thereby generating two Fab (fragment antigen binding) fragments and one Fc (fragment crystallizable) fragment (Figure 1.3). Each Fab fragment contains one of the two identical combining sites of the immunoglobulin.

Pepsin cleaves the heavy chain of IgG at the C-terminal side of the disulphide bridges, generating a single bivalent $F(ab')_2$ fragment containing the two combining sites. The $F(ab')_2$ fragment can be dissociated into univalent Fab′ after reduction of the disulphide bridges. Under certain conditions, pepsin digestion can yield Fv fragments which consist

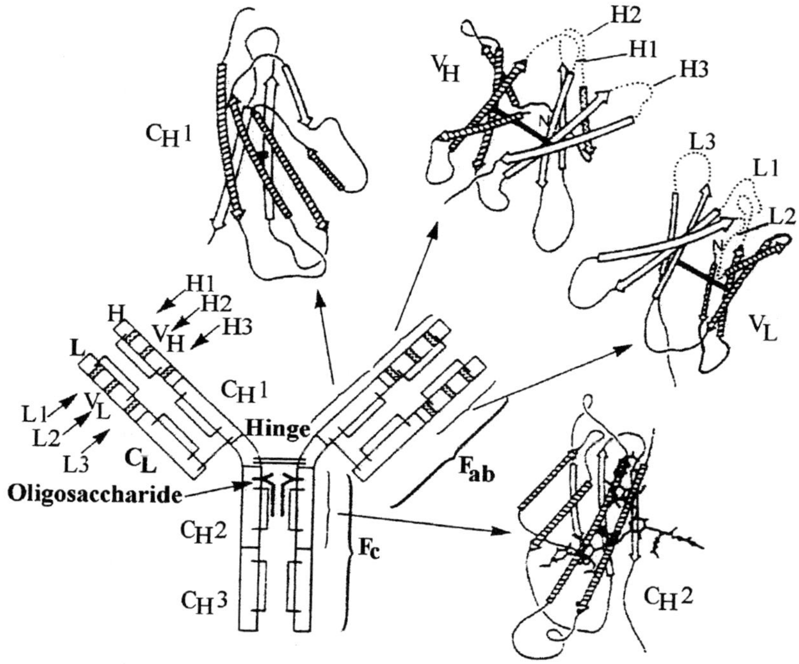

Figure 1.3 Schematic representation of an antibody molecule (human subclass IgG1). The homologous domains within the heavy (H) and light (L) chains are indicated, and the hypervariable segments within the variable (V) regions are shown. Inter- and intra-chain disulphide bridges are indicated by solid lines. The folding of the polypeptide chain within individual domains is also shown; the arrows represent strands of β-sheets, the solid bars are the intra-domain disulphide bridges, and the homologous four-stranded sheets of each domain are cross-hatched. The location of the hypervariable loop regions (H1–3, L1–3) may be seen within the V domains. An *N*-linked oligosaccharide chain covers part of the four-stranded face of each C_{H2} domain (from Sutton, 1993, with permission).

of only the variable regions of the H and L chains and which retain therefore the antigen-binding capacity of the parent immunoglobulin. Fv fragments are nowadays easily obtained by recombinant DNA technology (Borrebaeck, 1995).

Single-chain variable fragments (scFv) consist of the V_L and V_H domains of immunoglobulins linked by a flexible peptide (Plückthun, 1994). The scFv possesses the same binding capacity as the parental immunoglobulin.

The V_L and V_H domains of immunoglobulins associate non-covalently to form a structure that places the six CDR loops close together, thereby creating the antibody combining site or paratope. Since paratopes are built up from residues located on most or all of the CDRs, they clearly correspond to discontinuous structures (Figure 1.1). However, it has been shown in recent years that certain peptides corresponding to a single CDR sequence can bind the antigen with similar specificity as the whole antibody, albeit with lower affinity (Laune *et al.*, 1997). This is reminiscent of the situation found with certain continuous epitopes that are actually part of a more complex discontinuous epitope. By analogy, CDR peptides that possess antigen-binding activity may be called continuous paratopes (Van Regenmortel, 1998a).

Our knowledge of the structural basis of epitope–paratope interaction has been derived mainly from X-ray diffraction studies of complexes of Mab fragments with their protein antigens (Padlan, 1994; Braden and Poljak, 1995). Usually an area of about 700–900 $Å^2$ of the antigen surface, comprising atoms from 15 to 22 amino acid residues, is found to be in contact with an equivalent paratope surface. It seems that many of the epitope and paratope residues said to be in contact do not contribute to the energy of interaction. Mutagenesis studies and chemical replacement studies have shown that only three to five residues of the epitope defined in structural terms contribute significantly to the binding energy (Kelley, 1994).

Since an individual paratope is made up of not more than 15–20 residues, this means that about two-thirds of the approximately 50 hypervariable residues of the antibody do not participate in the individual interaction with a particular complementary epitope. These 'unused' residues of the six CDRs are available for binding to other epitopes that bear little or no structural resemblance to the first epitope. This is the reason why antibodies are multispecific and explains why the relation between an antibody and its antigen is never of an exclusive nature. A single immunoglobulin molecule could harbour for instance two totally independent subsites representing two paratopes for unrelated epitopes (Battacharjee and Glaudemans, 1978). A more likely finding is that the different subsites present in the immunoglobulin-binding pocket partly overlap, in which case binding to one epitope will prevent a second epitope from being accommodated at a nearby location.

In addition to recognizing the antigen against which it was elicited, an antibody will always bind to a variety of related or unrelated antigens (Roberts *et al.*, 1993). Usually, the antibody will bind to its homologous antigen with higher affinity than to most heterologous antigens. However, it is not rare to find heterologous antigens to which the antibody binds more strongly than to the homologous immunogen, a phenomenon known as heterospecific or heteroclitic binding (Underwood, 1985; Van Regenmortel, 1994). Heterospecificity can be of considerable practical use when attempts are made to obtain from a single hybridoma fusion experiment a set of Mabs specific for the individual members of a family of related proteins (Frison and Stace-Smith, 1992).

1.4 Monoclonal antibodies

Mabs are increasingly replacing polyclonal antibodies in most immunochemical studies in plant science, including plant pathology. A major advantage of Mabs over polyclonal reagents is the possibility of obtaining practically unlimited quantities of the same antibody to almost any epitope of interest, even when only impure antigen is available for immunization. The development of hybridoma technology has clarified the structural basis of epitope–paratope interactions and has removed the difficulties of interpretation encountered in earlier studies when antigens were analysed by means of heterogeneous antisera. Epitope mapping by means of a panel of Mabs has clarified the nature of the cross-reactions that occur between antigens. As shown in Figure 1.4, the cross-reaction between two antigens may be due to the shared reactivity that occurs when a Mab recognizes the same epitope a in the two antigens. It may also be due to a genuine cross-reactivity arising from the ability of a Mab to recognize two related epitopes such as c and e in Figure 1.4. The question is often asked whether a Mab or a polyclonal antiserum should be considered the more specific reagent. The question is actually meaningless, unless it includes a reference about what is to be differentiated. If the aim is to distinguish between antigen 1 and 2 (Figure 1.4), Mab anti-a would be considered non-specific, whereas Mabs anti-b and anti-d would be considered highly specific since they do not

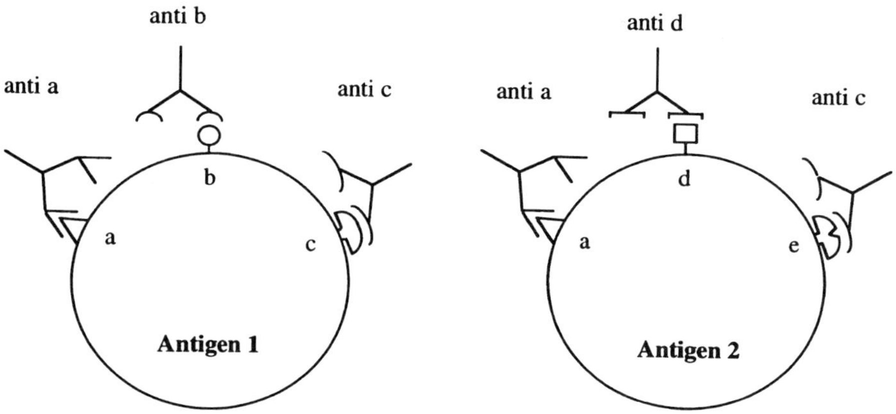

Figure 1.4 Potential cross-reactions between two antigens 1 and 2. Since epitope a is present on both antigens, Mab anti-a will react in an identical fashion with both antigens. Mabs anti-b and anti-d recognize unrelated epitopes on the two antigens and will show no cross-reactivity. Mab anti-c reacts strongly with the homologous epitope c and cross-reacts weakly with heterologous epitope e.

cross-react at all. For discriminating between the two antigens, a polyclonal antiserum containing a mixture of antibodies anti-a, -b and -c would be more specific than Mab anti-a, but less specific than Mab anti-b. It is in fact misleading to refer to a Mab as being specific for such entities as a protein antigen or a viral strain or serotype since an antibody can only be specific for an epitope. Part of the confusion surrounding the concept of immunochemical specificity originates from the lack of differentiation between epitope and multi-epitopic antigen (Van Regenmortel, 1998b). Reagents tend to be called specific if they achieve the type of discrimination that the investigator wishes to make. It is, therefore, more meaningful to speak of the discrimination potential of a Mab than of its specificity.

Procedures for obtaining Mabs have been described in many texts (Campbell, 1991; Goding, 1996) and reviews devoted to the use of Mabs in plant pathology (Hampton *et al.*, 1990; Van Regenmortel and Dubs, 1993) and in plant science (Wang, 1986) are available.

1.5 Antigen–antibody interactions

All antigen–antibody interactions are reversible and the energy of dissociation is usually higher than the energy of association because of the formation of additional secondary bonds. Because of hysteresis, more energy is needed to dissociate most antigen–antibody bonds than is required to prevent their formation (Van Oss and Van Regenmortel, 1994). The relationship between the dissociation rate constant (k_d) and the half-life of an interaction is presented in Table 1.2. Half-lives are relevant for predicting the outcome of an immunoassay. If a Mab has a k_d value of 10^{-2} s^{-1}, it will be of little use in a solid-phase immunoassay, since the antibody will become dissociated during the washing step of the assay. The determination of rate constants with biosensors is discussed in Chapter 5 of this book. The physico-chemical principles that control the outcome of an immunoassay have been discussed elsewhere (Van Regenmortel, 1997).

Table 1.2 Relationship between dissociation rate constant (k_d) and half-life ($t_{1/2}$) of an interaction

k_d (s^{-1})	$t_{1/2}$
10^{-6}	8 days
5×10^{-6}	1.6 days
10^{-5}	19.2 hours
5×10^{-5}	3.8 hours
10^{-4}	1.9 hours
5×10^{-4}	23 min
10^{-3}	11.5 min
5×10^{-3}	2.3 min
10^{-2}	70 s
10^{-1}	7 s

1.6 Antibodies expressed in plants

Among the various recombinant proteins that can be expressed in transgenic plants, antibodies offer the widest range of applications (Hiatt *et al.*, 1992; Hiatt and Ma, 1993). Plants are capable of synthesizing virtually every kind of antibody molecule, ranging from scFv fragments (Owen *et al.*, 1992; Firek *et al.*, 1993; Tavladoraki *et al.*, 1993; Artsaenko *et al.*, 1995; Fiedler and Conrad, 1995; Fecker *et al.*, 1996; Rosso *et al.*, 1996; Schouten *et al.*, 1996; Fiedler *et al.*, 1997; Phillips *et al.*, 1997; Schouten *et al.*, 1997) to Fab fragments (De Neve *et al.*, 1993), to full-length antibody molecules (Hiatt *et al.*, 1989; Düring *et al.*, 1990; Hein *et al.*, 1991; Stieger *et al.*, 1991; Van Engelen *et al.*, 1994; Voss *et al.*, 1995; Baum *et al.*, 1996) and even multimeric antibodies (Ma *et al.*, 1994, 1995). Hybrid immunoglobulin A–G heavy chains have also been produced (Ma *et al.*, 1994, 1995).

Plant systems are also being considered as an alternative to animal cell culture for the commercial production of Mabs. Large-scale production of Mabs using hybridoma cell culture requires nutrient media containing complex and expensive protein additives which interfere with purification of antibody from the culture broth (Wongsamuth and Doran, 1996). In comparison, the large-scale production of antibodies in plants is relatively inexpensive. It has been estimated that 1 g of recombinant antibody can be produced in leaves of a plant crop cultivated in an area of about 4 m^2 (Cockburn *et al.*, 1994). Compared with current prices for Mabs which range from US$2000 to $5000 per gram, the direct cost associated with producing 1 kg of antibody using plants has been estimated to be only about US$100 (Hiatt, 1990).

Recent advances in plant genetic engineering have made it possible to express antibodies and antibody fragments in higher plants (Hiatt *et al.*, 1989; Düring *et al.*, 1990; Owen *et al.*, 1992; Tavladoraki *et al.*, 1993; Conrad and Fiedler, 1994; Van Engelen *et al.*, 1994; Artsaenko *et al.*, 1995; Fiedler and Conrad, 1995; Voss *et al.*, 1995; Baum *et al.*, 1996; Phillips *et al.*, 1997), in callus (Hein *et al.*, 1991; De Neve *et al.*, 1993), in plant cell suspension cultures (Firek *et al.*, 1993; Van Engelen *et al.*, 1994; Fecker *et al.*, 1996), in protoplasts (Hein *et al.*, 1991; Rosso *et al.*, 1996; Schouten *et al.*, 1997) as well as in hairy root cultures (Wongsamuth and Doran, 1996).

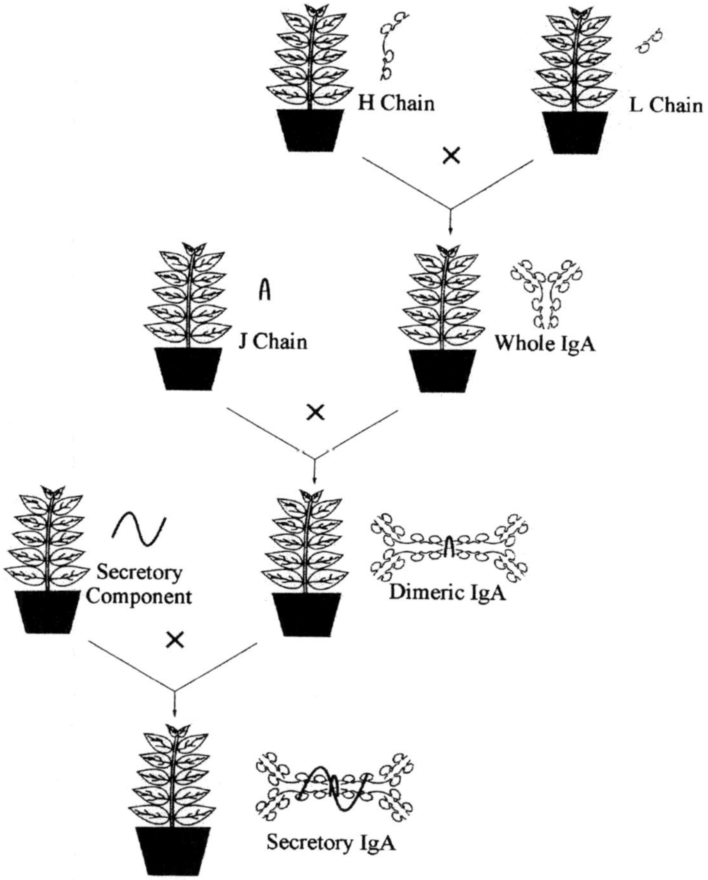

Figure 1.5 Cross-pollination strategy for combining genes encoding the four separate components of a secretory, polymeric antibody in an individual tobacco (*Nicotiana tabacum*) plant. Components and complexes synthesized are shown to the right of the plants that produce them (from Whitelam and Cockburn, 1996).

Although it is possible to transform and regenerate a wide variety of plants, production of plantibodies has so far only been carried out in tobacco (*Nicotiana tabacum* and *Nicotiana benthamiana*), *Arabidopsis thaliana*, *Acetabularia mediterranea* and *Solanum tuberosum*.

Plant cells can be propagated as callus, liquid-suspension cultures, or protoplasts, i.e., cells from which the walls have been removed, and whole plants can be regenerated from all three systems. Sexual crossing of transgenic plants can be used to introduce additional genes into the progeny and the genes required for complex oligomeric compounds can be accumulated by a series of crosses (Figure 1.5) (Ma *et al*., 1995).

1.6.1 Introduction of antibody genes in plants

Different techniques have been used to introduce antibody genes in plants. The most convenient is *Agrobacterium* transformation. Microinjection of cDNA into isolated nuclei before reimplantation into an enucleated cell has been achieved by Stieger *et al.* (1991)

and transient expression of antibodies in tobacco leaf protoplasts has been performed (Van Engelen *et al.*, 1994; Rosso *et al.*, 1996; Schouten *et al.*, 1996).

The gene for each immunoglobulin chain can be introduced separately in different plants. Subsequent cross-pollination of individual heavy- and light-chain-expressing plants makes it possible to bring the two genes together in one progeny (Figure 1.5) (Hiatt *et al.*, 1989; Ma *et al.*, 1994). The use of cross-pollination of individually transformed plants expressing antibody light or heavy chains involves two generations of plants for developing an antibody-producing plant and may not always be possible, as in the case of vegetatively propagated species. The yield of recombinant antibody is consistently high by this technique (1–5 per cent of total soluble protein; TSP). Another approach consists of using double-transformation techniques to introduce both the heavy- and light-chain genes into the same plant cell simultaneously (De Neve *et al.*, 1993). The time required to produce an antibody-secreting plant is reduced but this method increases the time needed for screening because of differences in the level of expression of the two genes. The yield of antibody was found to be lower by this technique and amounted to only 0.055 per cent of TSP (De Neve *et al.*, 1993). An alternative strategy is to introduce the light- and heavy-chain genes on a single sequence of DNA (Düring *et al.*, 1990; Van Engelen *et al.*, 1994). This is the most rapid technique and avoids problems of segregation in the progeny. The promoter and terminator must be selected in an appropriate manner to coordinate expression of the two transgenes. Up to 1.1 per cent of TSP of such a transgenic plant may consist of plantibodies (Van Engelen *et al.*, 1994).

For ubiquitous expression in plants, transcription of antibody genes can be controlled by the 35S promoter of *Cauliflower mosaic virus* (genus *Geminivirus*, CaMV) (Hiatt *et al.*, 1989; Hein *et al.*, 1991; De Neve *et al.*, 1993; Firek *et al.*, 1993; Tavladoraki *et al.*, 1993; Ma *et al.*, 1994; Baum *et al.*, 1996), by the enhanced CaMV 35S promoter (Van Engelen *et al.*, 1994; Voss *et al.*, 1995) linked with the Ω-leader region of *Tobacco mosaic virus* (genus *Tobamovirus*, TMV) (Fecker *et al.*, 1996) or with the 38 bp RNA4 untranslated leader of *Alfalfa mosaic virus* (genus *Alfamovirus*, AlMV) (Schouten *et al.*, 1996). The dual promoter from the octopine Ti-plasmid from *A. tumefaciens* has been used to express antibody L chain concomitantly with the promoter of the nopaline synthetase gene for expression of antibody H chain (Van Engelen *et al.*, 1994). The TR2′ dual promoter was used for ubiquitous expression and high level expression (Düring *et al.*, 1990; Stieger *et al.*, 1991; Rosso *et al.*, 1996).

Seed-specific legumin LeB4 promoter (Fiedler and Conrad, 1995) and the unknown USP seed protein promoter from *Vicia faba* (Fiedler *et al.*, 1997; Phillips *et al.*, 1997) were used for expression in storage organ specific (tobacco seeds) for long-term storage of the foreign protein in the harvested plant material. When put under control of the seed-specific USP promoter, expression levels of scFv antibodies reached up to 3–4 per cent of TSP in tobacco seeds (Phillips *et al.*, 1997).

1.6.2 *Assembly of antibodies in plants*

Several groups have reported the production in plants of fragmented or full-length antibodies that are functional in terms of antigen recognition and binding. An important advantage of plants over other recombinant expression systems is the ability to assemble full-length heavy chains with light chains to form full-length active antibody. In most cases, individual light and heavy chains or misfolded immunoglobulin molecules are not functional.

In mammalian cells, the assembly mechanism is only partially understood. Immunoglobulin light and heavy chains are synthesized as precursor proteins and signal sequences direct the translocation into the lumen of the endoplasmic reticulum (ER). Within the ER, there is cleavage of the signal peptides and at least two stress proteins functioning as chaperones bind to assembled H and L chains and direct subsequent folding and assembly.

In plants, passage through the ER is also required for correct assembly of full-length antibodies and Fab fragments. In the absence of a signal peptide linked to either L or H chain gene, assembly of antibody does not take place (Hiatt *et al.*, 1989). However, functional antibodies have been assembled inside the cytoplasm (Stieger *et al.*, 1991). Chimeric immunoglobulin genes lacking the signal sequence coding for the signal peptide responsible for docking the mRNA–ribosome complex to the ER have been injected into the nucleus of the alga *Acetabularia mediterranea*. Antibodies were formed with correctly folded antigenic binding sites, indicating that the concentration of reducing agents in the cytoplasm needed to prevent formation of disulphide bonds was not very high in this organism (Stieger *et al.*, 1991).

Signal sequences from a variety of sources have been used successfully. They include non-plant signal sequences such as mouse immunoglobulin sequences (Hiatt *et al.*, 1989; Ma *et al.*, 1994; Van Engelen *et al.*, 1994; Voss *et al.*, 1995), yeast pre-prosequence of α-mating factor (Hein *et al.*, 1991) and two plant signal sequences such as barley aleurone α-amylase (Düring *et al.*, 1990) and 2S2 protein of *Arabidopsis thaliana* (De Neve *et al.*, 1993). However, unexpected targeting sometimes occurs as reported in the case of chloroplasts (Düring *et al.*, 1990) and the nucleolus of meristematic tissue (De Neve *et al.*, 1993).

Proteolytic processing of the signal sequence occurs within the plant ER. Plant chaperones homologous to mammalian chaperones have been identified in the lumen of the ER, such as a BIP protein, which is a key molecule in the protein maturation and transport machinery of the endomembrane system (Denecke *et al.*, 1991) and HSP90, another chaperone protein cloned from barley leaf, which is involved in folding and assembly of newly synthesized proteins in the ER (Walther-Larsen *et al.*, 1993). Assembly of a stable active antibody involves formation of intramolecular disulphide bonds which is greatly facilitated in the ER because of the presence of disulphide isomerase (Düring *et al.*, 1990).

Plantibodies present an altered glycosylation patterns and lack N-acetyl neuraminic acid, which is the predominant terminal sugar residue in mammalian complex glycans (Franken *et al.*, 1997). N-linked glycosylation of proteins in plants is similar to the glycosylation process in mammals with glucosyl terminal residues consisting of xylose, fucose, N-acetylglucosamine, mannose or glucose (Sturm *et al.*, 1987). A core high-mannose oligosaccharide is attached to the asparagine contained within the canonical Asn–X–Ser/Thr sequence. This occurs in the ER and can be modified in the Golgi apparatus where terminal sugars are attached (Sturm *et al.*, 1987). Both plant and murine glycans are resistant to endoglycosidase H, which is characteristic of complex carbohydrates processed in the Golgi apparatus, indicating that the transgenic antibody is processed in a similar fashion to complex mammalian glycoproteins (Kijimoto-Ochiai *et al.*, 1989).

ScFv fragment antibodies can be assembled in the cytoplasm of plant cells without processing through the ER lumen. The quantity of scFv obtained without a signal sequence is very low and is usually 0.06–0.1 per cent of TSP (Owen *et al.*, 1992; Tavladoraki *et al.*, 1993). A plant expression system where the scFv-proteins are targeted to the ER provides a higher accumulation level of active single-chain Fv antibodies (Fiedler *et al.*,

1997). This permits a tenfold increase of the quantity of produced antibody (Firek *et al.*, 1993). Different signal sequences have been tested. The most commonly employed one is the mouse immunoglobulin signal sequence (Rosso *et al.*, 1996; Schouten *et al.*, 1996). The nature of the signal sequence is more important for scFv fragments than for Fab fragments or full-length immunoglobulins. The accumulation of scFv fragments varies, depending on the antibody and linker sequences. In one study, no scFv could be detected when mouse immunoglobulin was used as the signal sequence, apparently because the secreted scFv was degraded (Rosso *et al.*, 1996). Signal sequences of plants that have been used successfully are the bean phytohaemagglutinin (PHA) signal sequence (Fecker *et al.*, 1996), the tobacco pathogenesis-related protein PR1a signal sequence (Firek *et al.*, 1993), and the legumin B4 signal sequence of *Vicia faba* (Artsaenko *et al.*, 1995; Fiedler and Conrad, 1995).

To stabilize the cytosolic expression of scFv in plants, the C-terminal KDEL sequence, which is an ER retention signal, has been used to target antibody away from the reducing cytosol to the potentially more favourable environment of the ER. The retention of antibodies in the ER resulted in a 100-fold increase in the amount of detectable scFv antibodies (Schouten *et al.*, 1996). Moreover, changing from KDEL to KDEI did not affect the enhanced protein expression level (Schouten *et al.*, 1997).

Correctly folded full-length antibodies are secreted through the plasma membrane of the cell and, because of their size, are retained in the apoplast between the plasma membrane and the cell wall. Plant cells have the capacity of retaining antibodies in a hemipermeable membrane. However, antibodies are secreted through both the plasma membrane and the cell wall of callus cells. This is somewhat surprising since the permeability of plant cells to macromolecules is thought to be very limited. Evaluation of macromolecule migration through plant cell walls from a variety of sources has demonstrated an exclusion limit equivalent to 20 kDa globular proteins (Carpita *et al.*, 1979). Secretion of antibodies (M_r = 150 000) through the cell wall of callus suspension cells may reflect the presence of a small population of large pores in the plant cell wall which allows passage of large proteins or extracellular carbohydrates (Hein *et al.*, 1991). The full-size antibody directed against a cutinase produced by the fungus *Botrytis cinerea* was partially retained by the wall of suspension cells. Once in the apoplast, antibodies appear to be subject to proteolytic processing as indicated by the occurrence of a 110 kDa $F(ab')_2$-like antibody fragment in a cell suspension which can freely diffuse through cell walls (Van Engelen *et al.*, 1994). Moreover, the scFv antibodies were not secreted into the culture fluids of suspension cultures, but were retained in the cells (Fecker *et al.*, 1996).

1.6.3 *Plantibodies for* ex situ *applications*

The successful production of an intact, recombinant antibody catalysing the hydrolysis of certain carboxylic esters in tobacco plants laid the foundation for current work aimed at the development of plants as antibody expression systems (Hiatt *et al.*, 1989).

A potential area of application is bioremediation since plants can be produced that bind pollutants such as polychlorinated biphenyls (PCBs) and herbicides from contaminated water and soil (Hiatt, 1990).

Since functional secretory antibodies can be engineered, immunotherapy is one of the many potential uses of plantibodies (Ma *et al.*, 1989). Antibody molecules of potential therapeutic importance have been produced in plants (Ma *et al.*, 1995), such as the multimeric secretory IgA which is the predominant form of immunoglobulin in secretions from mucosal surfaces. A series of sexual crosses of four transgenic *Nicotiana tabacum* plants was performed in order to generate plants in which all four polypeptides (L and H chains of immunoglobulin, a linker peptide and a secretory signal peptide) were expressed simultaneously. In plants, a single cell is able to assemble the secretory antibody (Ma and Hein, 1995a, 1995b).

1.6.4 *Plantibodies for* in situ *applications*

Plant growth and development is controlled by a limited number of low molecular weight hormones such as indoleacetic acid, ethylene, benzylaminopurine and a variety of more complex organic molecules (Hiatt, 1990). By expressing within the plant cell monoclonal antibodies that recognize these hormones, it may be possible to evaluate developmental and metabolic events (Hiatt, 1990). Plantibodies are powerful tools for modulating metabolic pathways in plants. Antibodies can be used to bind to and therefore modify the activity of one or more of the plant components (Owen *et al.*, 1992). In the first of these studies, a cDNA was expressed that encoded a cytoplasmically targeted scFv derived from a monoclonal antibody directed against a conserved epitope of the regulatory photoreceptor, phytochrome A (Owen *et al.*, 1992). Transgenic tobacco plants expressing the scFv gene accumulated a low level of the scFv protein that was found to display antigen-binding activity following extraction. Phytochrome-dependent promotion of germination and photocontrol of hypocotyl elongation were both aberrant in transgenic plants, indicating that the scFv protein was functional *in vivo* (Owen *et al.*, 1992). Plantibodies allow modulation of activities of molecules not accessible to antisense technology, such as the plant hormone abscisic acid in transgenic tobacco (Artsaenko *et al.*, 1994, 1995). By introducing a cDNA encoding a scFv specific for the plant hormone abscisic acid (ABA) into transgenic tobacco, a wilty phenotype was created, similar to that of ABA-deficient mutants (Artsaenko *et al.*, 1995). The mechanism by which *in situ* antibody expression causes ABA deficiency symptoms is not known. One possibility is that the ER-localized antibody leads to hormone sequestration in this compartment, thereby reducing the concentration of free ABA elsewhere in the cell (Artsaenko *et al.*, 1995). Seeds of transgenic plants expressing anti-ABA scFv germinated precociously if removed from capsules during development but were incapable of germination after drying. This is the first case of seed-specific immunomodulation which results in a switch from the seed maturation programme to the germination programme (Phillips *et al.*, 1997).

1.6.5 *Plantibodies for controlling pathogens*

A wide range of viruses infect susceptible crops with devastating effects. Molecular biological techniques have the potential to accelerate the introduction of virus-resistant plant phenotypes. Sequences encoding viral proteins such as coat protein, movement protein or replicase as well as antisense, satellite and defective interfering RNA molecules have been expressed in transgenic plants. Engineered antibodies have been transfected into plant cells to endow plants with new forms of defence against diseases. Expression of an intracellular antibody that binds antigens essential for pathogenesis may reduce the

symptoms of the infection by reducing the functional titre. The advantage of this strategy is twofold: first, it would not require the isolation of genes involved in synthesis of the target antigen; and second, pools of antigens which may be localized in subcellular compartments could be specifically targeted, leaving other pools unaffected (Hiatt, 1990). Crops with built-in resistance to pests, pathogens or viral infections are currently being developed using recombinant antibody techniques.

Tavladoraki *et al.* (1993) first demonstrated that transgenic *Nicotiana benthamiana* plants can be protected from viral attack by the constitutive cytoplasmic expresssion of an engineered antibody (scFv) derived from a mouse immunized against the *Artichoke mottled crinkle virus* (genus *Tombusvirus*, AMCV). The resulting transgenic lines showed a lower virus accumulation, a highly reduced incidence of infection and a significant delay in the appearance of viral symptoms compared with untransformed plants. It was the first demonstration of a plant phenotype with an attenuation of viral infection derived from a constitutively expressed, virus-specific antibody.

Mabs to *Beet necrotic yellow vein virus* (genus *Benyvirus*, BNYVV) coat protein (CP) or to the 25-kDa non-structural protein (P25) were cloned to produce scFv antibodies in *Nicotiana benthamiana*. Detectable amounts were found in plants when the scFv were targeted to the secretory pathway by means of the PHA signal sequence. No transgenic plants were regenerated when using P25-specific scFv. However, several plants were obtained with the CP-specific scFv, but whether BNYVV CP-specific scFv interfere with the activities of the virus in infected plants was not determined (Fecker *et al.*, 1996). Production of scFv in plants can interfere with development of the plant. A total of 25 transgenic plants obtained by expressing scFv-specific CP BNYVV developed normally whereas five transgenic plants expressing scFv-specific P25 BNYVV grew poorly and did not form roots (Fecker *et al.*, 1996; see also Chapter 10 of this book).

A neotope-specific full-length antibody which links epitopes present only on the surface of intact tobacco mosaic virions was used to produce transgenic plants expressing a full-length antibody (Voss *et al.*, 1995). Upon infection with TMV, F_1 plants producing plantibodies showed a reduction in necrotic lesion numbers. The maximal reduction was 70 per cent and the reduction level was directly correlated to the amount of secreted antibody. The total amount of assembled full-size antibodies and the expression levels were regulated by the H chain transcript (Voss *et al.*, 1995). Other workers showed reduction of necrotic local lesions in plants expressing a scFv specific for a TMV neotope. The most drastic reduction of necrotic local lesion numbers upon virus infection was observed in F_1 plants expressing scFv fragments in the cytosol. The number of local lesions could be reduced by more than 90 per cent, despite the fact that expression levels for functional scFv antibodies were very low compared with scFv targeted to the secretory pathway. Expression of TMV-specific scFv in the cytosol conferred virus resistance in these transgenic plants. Inoculation of protoplasts with TMV-RNA revealed that viral RNA replication was affected by scFv expression (Schillberg *et al.*, 1997).

A monoclonal antibody specific to stylet secretions of *Meloidogyne incognita*, a plant parasite, was used to produce full-length antibody molecules in tobacco. Although plantibodies were expressed in leaves, stems, roots and galls, their expression had no influence on root-knot nematode parasitism of transgenic plants, suggesting that antibodies and nematode stylet secretions might be localized in different cellular compartments and could not interact (Baum *et al.*, 1996). Rosso *et al.* (1996) developed a scFv strategy using the same monoclonal antibody. The scFv fragments were expressed in transgenic tobacco cells but whether scFv and nematode secretions interacted was not determined (Rosso *et al.*, 1996; see also Chapter 11 of this book).

References

ARTSAENKO, O., WEILER, E.W., MÜNTZ, K. and CONRAD, U., 1994, Construction and functional characterization of a single chain Fv antibody binding to the plant hormone abscisic acid, *Journal of Plant Physiology*, **144**, 427–429.

ARTSAENKO, O., PEISKER, M., ZUR NIEDEN, U., FIEDLER, U., WEILER, E.W., MÜNTZ, K. and CONRAD, U., 1995, Expression of a single-chain Fv antibody against abscisic acid creates a wilty phenotype in transgenic tobacco, *The Plant Journal*, **8**, 745–750.

BATTACHARJEE, A.K. and GLAUDEMANS, C.P.J., 1978, Dual binding specificities in MOPC 384 and 870 murine myeloma immunoglobulins, *Journal of Immunology*, **120**, 411–413.

BAUM, T.J., HIATT, A., PARROTT, W.A., PRATT, L.H. and HUSSEY, R.S., 1996, Expression in tobacco of a functional monoclonal antibody specific to stylet secretions of the root-knot nematode, *Molecular Plant–Microbe Interactions*, **9**, 382–387.

BENJAMIN, D.C., BERZOFSKY, J.A., EAST, I.J., GURD, F.R.N., HANNUM, C., LEACH, S.J., *et al.*, 1984, The antigenic structure of proteins: a reappraisal, *Annual Review of Immunology*, **2**, 67–101.

BORREBAECK, C.A.K. (ed.), 1995, *Antibody Engineering* (second edition), Oxford University Press, Oxford.

BRADEN, B.C. and POLJAK, R.J., 1995, Structural features of the reactions between antibodies and protein antigens, *FASEB Journal*, **9**, 9–16.

CAMPBELL, A.M., 1991, *Monoclonal Antibody and Immunosensor Technology*, Elsevier, Amsterdam.

CARPITA, N., SABULARSE, D., MONTEZINOS, D. and DELMER, D.P., 1979, Determination of the pore size of cell walls of living plant cells, *Science*, **205**, 1144–1147.

COCKBURN, W., WHITELAM, G.C., OWEN, M., GANDECHA, A., RIDDELL, K. and LAVERTY, E., 1994, Exploration of the potential plant crops for the large-scale production of recombinant antibodies, pp. 19–20, Second Conference on Industrial Immunology, Brighton, UK, July.

CONRAD, U. and FIEDLER, U., 1994, Expression of engineered antibodies in plant cells, *Plant Molecular Biology*, **26**, 1023–1030.

DENECKE, J., GOLDMAN, M.H., DEMOLDER, J., SEURINCK, J. and BOTTERMAN, J., 1991, The tobacco luminal binding protein is encoded by a multigene family, *Plant Cell*, **3**, 1025–1035.

DE NEVE, M., DE LOOSE, M., JACOBS, A., VAN HOUDT, H., KALUZA, B., WEIDLE, U., *et al.*, 1993, Assembly of an antibody and its derived antibody fragment in *Nicotiana* and *Arabidopsis, Transgenic Research*, **2**, 227–237.

DEOM, C.M., LAPIDOT, M. and BEACHY, R.N., 1992, Plant virus movement protein, *Cell*, **69**, 221–224.

DIMMOCK, N.J., 1993, *Neutralization of Animal Viruses*, Springer-Verlag, Berlin, p. 149.

DORE, I., RUHLMANN, P., OUDET, P., CAHOON, M., CASPAR, D.L.D. and VAN REGENMORTEL, M.H.V., 1990, Polarity of binding of monoclonal antibodies to tobacco mosaic virus rods and stacked disks, *Virology*, **176**, 25–29.

DÜRING, K., HIPPE, S., KREUZALER, F. and SCHELL, J., 1990, Synthesis and self-assembly of a functional monoclonal antibody in transgenic *Nicotiana tabacum*, *Plant Molecular Biology*, **15**, 281–293.

FECKER, L.F., KAUFMANN, A., COMMANDEUR, U., COMMANDEUR, J., KOENIG, R. and BURGERMEISTER, W., 1996, Expression of single-chain antibody fragments (scFv) specific for beet necrotic yellow vein virus coat protein or 25 kDa protein in *Escherichia coli* and *Nicotiana benthamiana*, *Plant Molecular Biology*, **32**, 979–986.

FIEDLER, U. and CONRAD, U., 1995, High-level production and long-term storage of engineered antibodies in transgenic tobacco seeds, *Bio/Technology*, **13**, 1090–1093.

FIEDLER, U., PHILLIPS, J., ARTSAENKO, O. and CONRAD, U., 1997, Optimization of scFv antibody production in transgenic plants, *Immunotechnology*, **3**, 205–216.

FIREK, S., DRAPER, J., OWEN, M.R.L., GANDECHA, A., COCKBURN, B. and WHITELAM, G.C., 1993, Secretion of a functional single-chain Fv protein in transgenic tobacco plants and cell suspension cultures, *Plant Molecular Biology*, **23**, 861–870.

FRANKEN, E., TEUSCHEL, U. and HAIN, R., 1997, Recombinant proteins from transgenic plants, *Current Opinion in Biotechnology*, **8**, 411–416.

FRISON, E.A. and STACE-SMITH, R., 1992, Cross-reacting and heterospecific monoclonal antibodies produced against arabis mosaic nepovirus, *Journal of General Virology*, **73**, 2525–2530.

GODING, J.W., 1996, *Monoclonal Antibodies: Principles and Practice* (third edition), Academic Press Inc., San Diego.

HAMPTON, R., BALL, E. and DEBOER, S., 1990, *Serological Methods for Detection and Identification of Viral and Bacterial Plant Pathogens*, American Phytopathological Society Press, St Paul, Minnesota.

HEIN, M.B., TANG, Y., MCLEOD, D.A., JANDA, K.D. and HIATT, A., 1991, Evaluation of immunoglobulins from plant cells, *Biotechnology Progress*, **7**, 455–461.

HIATT, A., 1990, Antibodies produced in plants, *Nature*, **344**, 469–470.

HIATT, A. and MA, J.K., 1993, Characterization and applications of antibodies produced in plants, *International Reviews of Immunology*, **10**, 139–152.

HIATT, A., CAFFERKEY, R. and BOWDISK, K., 1989, Production of antibodies in transgenic plants, *Nature*, **342**, 76–78.

HIATT, A., TANG, Y., WEISER, W. and HEIN, M.B., 1992, Assembly of antibodies and mutagenized variants in transgenic plants and plant cell cultures, *Genetic Engineering*, **14**, 49–64.

HULL, R., 1989, The movement of viruses in plants, *Annual Review of Phytopathology*, **27**, 213–240.

KELLEY, R.F., 1994, Thermodynamics of protein-protein interaction studied by using BIAcore and single-site mutagenesis, *Methods: A companion to Methods in Enzymology*, **6**, 111–120.

KIJIMOTO-OCHIAI, S., KATAGIRI, Y.U., HATAE, T. and OKUYAMA, H., 1989, Type analysis of the oligosaccharide chains on microheterogeneous components of bovine pancreatic DNAse by the lectin-nitrocellulose sheet method, *Biochemical Journal*, **257**, 43–50.

LAUNE, D., MOLINA, F., FERRIERES, G., MANI, J.C., COHEN, P., SIMON, D., *et al.*, 1997, Systematic exploration of the antigen binding activity of synthetic peptides isolated from the variable regions of immunoglobulins, *Journal of Biological Chemistry*, **272**, 30937–30944.

LESCAR, J., PELLEGRINI, M., SOUCHON, H., TELLO, D., POLJAK, R.J., PETERSON, N., *et al.*, 1995, Crystal structure of a cross-reaction complex between Fab F9.13.7 and guinea fowl lysozyme, *Journal of Biological Chemistry*, **270**, 18067–18076.

MA, J.K.C. and HEIN, M.B., 1995a, Immunotherapeutic potential of antibodies produced in plants, *Trends in Biotechnology*, **13**, 522–527.

MA, J.K.C. and HEIN, M.B., 1995b, Plant antibodies for immunotherapy, *Plant Physiology*, **109**, 341–346.

MA, J.K.C., HUNJAN, M., SMITH, R. and LERNER, T., 1989, Specificity of monoclonal antibodies in local passive immunization against *Streptoccoccus mutans, Infection and Immunity*, **58**, 3407–3414.

MA, J.K.C., LEHNER, T., STABILA, P., FUX, C.I. and HIATT, A., 1994, Assembly of monoclonal antibodies with IgG1 and IgA heavy chain domains in transgenic tobacco plants, *European Journal of Immunology*, **24**, 131–138.

MA, J.K.C., HIATT, A., HEIN, M., VINE, N.D., WANG, F., STABILA, P., *et al.*, 1995, Generation and assembly of secretory antibodies in plants, *Science*, **268**, 716–719.

MASSEYEFF, R.F., ALBERT, W.H. and STAINES, N.A., 1993, Methods of immunological analysis, in *Molecular Basis of Antigen–Antibody Reactions*, Vol. 1, pp. 66–114.

MORRIS, G.E., 1996, Epitope mapping by chemical fragmentation, in *Epitope Mapping Protocols*, Morris, G.E., ed., Humana Press Inc., 999 Riverview Dr, Ste 208, Totowa, NJ 07512, **66**, 121–127.

MUSHEGIAN, A.R. and KOONIN, E.V., 1993, Cell-to-cell movement of plant viruses. Insights from amino acid sequence comparisons of movement proteins and from analogies with cellular transport systems, *Archives of Virology*, **133**, 239–257.

NEZLIN, R., 1994, Immunoglobulin structure and function, in *Immunochemistry*, Van Oss, C.J. and Van Regenmortel, M.H.V., eds, Marcel Dekker, Inc., New York, pp. 3–45.

OWEN, M., GANDECHA, A., COCKBURN, B. and WHITELAM, G.C., 1992, Synthesis of a functional anti-phytochrome single-chain Fv protein in transgenic tobacco, *Bio/Technology*, **10**, 790–794.

PADLAN, E.A., 1994, Anatomy of the antibody molecule, *Molecular Immunology*, **31**, 169–217.

PHILLIPS, J., ARTSAENKO, O., FIEDLER, U., HORSTMANN, C., MOCK, H.P., MÜNTZ, K. and CONRAD, U., 1997, Seed-specific immunomodulation of abscisic acid activity induces a developmental switch, *EMBO Journal,* **16**, 4489–4496.

PLÜCKTHUN, A., 1994, Recombinant antibodies, in *Immunochemistry*, Van Oss, C.J. and Van Regenmortel, M.H.V., eds, Marcel Dekker, Inc., New York, pp. 201–236.

ROBERTS, V.A., GETZOFF, E.D. and TAINER, J.A., 1993, Structural basis of antigenic cross-reactivity, in *Structure of Antigens*, Van Regenmortel, M.H.V., ed., CRC Press, Boca Raton, pp. 31–53.

ROSSO, M.N., SCHOUTEN, A., ROOSIEN, J., BORST-VRENSSEN, T., HUSSEY, R.S., GOMMERS, F.J., *et al.*, 1996, Expression and functional characterization of a single chain Fv antibody directed against secretions involved in plant nematode infection process, *Biochemical and Biophysical Research Communications*, **220**, 255–263.

SAUNAL, H. and VAN REGENMORTEL, M.H.V., 1995, Kinetic and functional mapping of viral epitopes using biosensor technology, *Virology*, **213**, 462–471.

SCHILLBERG, S., ZIMMERMANN, S., LIAO, Y.-C., WILLMOTT, R. and FISCHER, R., 1997, Expression of TMV-specific recombinant antibodies leads to improved virus resistance in *Nicotiana tabacum*. Perspectives on Protein Engineering, unpublished communication.

SCHOUTEN, A., ROOSIEN, J., VAN ENGELEN, F.A., DE JONG, G.A.M., BORST-VRENSSEN, A.W.M., ZILVERENTANT, J.F., *et al.*, 1996, The C-terminal KDEL sequence increases the expression level of a single-chain antibody designed to be targeted to both the cytosol and the secretory pathway in transgenic tobacco, *Plant Molecular Biology*, **30**, 781–793.

SCHOUTEN, A., ROOSIEN, J., DE BOER, J.M., WILMINK, A., ROSSO, M.N., BOSCH, D., *et al.*, 1997, Improving scFV antibody expression levels in the plant cytosol, *FEBS Letters*, **415**, 235–241.

SHAW, J.G., PLASKITT, K.A. and WILSON, T.M.A., 1986, Evidence that tobacco mosaic virus particles disassemble cotranslationally *in vivo*, *Virology*, **148**, 326–336.

STIEGER, M., NEUHAUS, G., MOMMA, T., SCHELL, J. and KREUZALER, F., 1991, Self assembly of immunoglobulins in the cytoplasm of the alga *Acetabularia mediterranea*, *Plant Science*, **73**, 181–190.

STURM, A., KUIK, A.V., VLIEGENTHART, J.F.G. and CHRISPEELS, M.J., 1987, Structure, position, and biosynthesis of the high mannose and the complex oligosaccharide side chains of the bean storage protein phaseolin, *Journal of Biological Chemistry*, **262**, 13392–13403.

TAVLADORAKI, P., BENVENUTO, E., TRINCA, S., DE MARTINIS, D., CATTANEO, A. and GALEFFI, P., 1993, Trangenic plants expressing a functional single-chain Fv antibody are specifically protected from virus attack, *Nature*, **366**, 469–472.

UNDERWOOD, P.A., 1985, Theoretical considerations of the ability of monoclonal antibodies to detect antigenic differences between closely related variants, with particular reference to hetero-specific reactions, *Journal of Immunological Methods*, **85**, 295–307.

VAN ENGELEN, F.A., SCHOUTEN, A., MOLTHOFF, J.W., ROOSIEN, J., SALINAS, J., DIRKSE, W.G., *et al.*, 1994, Coordinate expression of antibody subunit genes yields high levels of functional antibodies in roots of transgenic tobacco, *Plant Molecular Biology*, **26**, 1701–1710.

VAN OSS, C.J. and VAN REGENMORTEL, M.H.V., 1994, in *Immunochemistry*, Van Oss, C.J. and Van Regenmortel, M.H.V., eds, Marcel Dekker, Inc., New York.

VAN REGENMORTEL, M.H.V., 1994, The recognition of proteins and peptides by antibodies, in *Immunochemistry*, Van Oss, C.J. and Van Regenmortel, M.H.V., eds., Marcel Dekker, New York, pp. 277–301.

Van Regenmortel, M.H.V., 1997, The antigen antibody reaction, in *Principles and Practice of Immunoassay* (2nd edition), Price, C.P. and Newman, D.J., eds, MacMillan Press, London, pp. 13–33.

Van Regenmortel, M.H.V., 1998a, Applications of synthetic peptides in the diagnosis of viral infections, *ASM News*, **64**, 332–338.

Van Regenmortel, M.H.V., 1998b, From absolute to exquisite specificity. Reflections on the fuzzy nature of species, specificity and antigenic sites, *Journal of Immunological Methods*, in press.

Van Regenmortel, M.H.V. and Dubs, M.C., 1993, Serological procedures, in *Diagnosis of Plant Virus Diseases*, Matthews, R.E.F., ed., CRC Press, pp. 159–214.

Van Regenmortel, M.H.V., Altschuh, D. and Zeder-Lutz, G., 1993, Tobacco mosaic virus: a model antigen to study virus–antibody interactions, *Biochimie*, **75**, 731–739.

Voss, A., Niersbach, M., Hain, R., Hirsch, H.J., Liao, Y.C., Kreuzaler, F. and Fischer, R., 1995, Reduced virus infectivity in *N. tabacum* secreting a TMV-specific full-size antibody, *Molecular Breeding*, **1**, 39–50.

Walther-Larsen, H., Brandt, J., Collinge, D.B. and Thordal-Christensen, H., 1993, A pathogen-induced gene of barley encodes a HSP 90 homologue showing striking similarity to vertebrate forms resident in the endoplasmic reticulum, *Plant Molecular Biology*, **21**, 1097–1108.

Wang, T.L. (ed.), 1986, *Immunology in Plant Science*, Cambridge University Press, Cambridge.

Whitelam, G.C. and Cockburn, W., 1996, Antibody expression in transgenic plants, *Trends in Plant Science*, **1**, 268–272.

Wilson, T.M.A., 1984, Cotranslational disassembly of tobacco mosaic virus *in vitro*, *Virology*, **137**, 255–265.

Wongsamuth, R. and Doran, P.M., 1996, Production of monoclonal antibodies by tobacco hairy roots, *Biotechnology and Bioengineering*, **54**, 401–415.

2

Cloning of Pre-existing Hybridomas – Production of Novel Diagnostic Tools for Plant Pathology

ANGELIKA ZIEGLER

2.1 Introduction

Immunoglobulin G (IgG, the most common serum globulin) is composed of two identical light and two identical heavy chains. The diversity in the variable domains of each of these chains is localized in highly variable loops. These are the complementarity determining regions (CDR) that form the antigen-binding pocket. The CDR are supported by a framework region of conserved amino acid sequence (Austyn and Wood, 1993; see also Chapter 1 of this book).

With increased knowledge of the organization of the immune system, the accumulation of sequence information on immunoglobulin genes, and also the invention of the polymerase chain reaction (PCR; Mullis *et al.*, 1986) it has become possible to clone genes encoding variable domains of antibodies and express them in *Escherichia coli*. With the help of translocation signals (for example, the pelB leader; Better *et al.*, 1988) the expressed proteins can be targeted to the periplasm, where the oxidizing environment allows proper folding of the antibodies and favours the formation of the intrachain disulphide bonds necessary for functional antibodies.

The antibody genes can be cloned in different formats, for example scFv (single-chain variable fragments) or Fab (fragment with antigen binding), and they can be genetically modified by combining them with peptide tags (for example, c-myc tag; Munro and Pelham, 1986; Ward *et al.*, 1989), signal proteins, toxins or enzymes as fusion proteins (see Chapter 4). Also, binding specificities can be modified for particular needs by the introduction of mutations in the CDRs or the framework (see Chapter 3).

Hybridoma cell lines are the product of the fusion of an antibody-producing B-cell and a myeloma cell. Each hybridoma secretes one particular antibody molecule called a monoclonal antibody (Mab; Köhler and Milstein, 1975). Mabs have provided tools for a wide range of immunological applications (see Chapter 1). RNA corresponding to the heavy and light chains of a Mab may be isolated from the hybridoma cells. The antibody variable regions of the hybridoma can be cloned, and the recombinant DNA representing the light and heavy chains can be stored as a stable source of the antibody genetic material (see Figure 2.1). This can be very helpful if, for example, a valuable, well characterized hybridoma line is unstable or does not grow well in cell culture. The recombinant antibodies

Mouse hybridoma cells

⇓

RNA extraction

⇓

cDNA synthesis

⇓

PCR amplification of heavy chain and light chain variable regions

⇓

Cloning into expression vector

⇓

scFv production in *E. coli*, subsequent purification and application

Figure 2.1 Outline of the cloning procedure.

can be produced in *E. coli* cheaply and quickly as standardized reagents, and there are no requirements for the use of animals or for large-scale cell culture. Kerschbaumer *et al.* (1997) have demonstrated the use of hybridoma-derived recombinant antibodies for the detection of *Beet necrotic yellow vein virus* (genus *Benyvirus*) and Terrada *et al.* (1998) have used these for the detection of *Citrus tristeza virus* (genus *Closterovirus*).

In this chapter, the cloning of variable antibody heavy and light chain genes from mouse hybridoma lines in the scFv format will be discussed. Examples will be given for the use of hybridoma-derived recombinant antibodies in the diagnosis of plant viruses.

2.2 Preliminary considerations

Freshly subcloned and productive hybridomas that secrete high affinity Mabs are the best for cloning purposes. The binding characteristics of a hybridoma cell line such as specificity, cross-reactivity with related antigens, or the usefulness of the antibody, for example, as an ELISA coating reagent (see Chapter 4) also have to be taken into account. The recombinant antibodies usually reflect the binding characteristics of their parent Mab. Therefore, if several Mabs are available for the same antigen, those best suited for the desired application should be chosen for cloning. Antibody secretion by the cell lines chosen should always be verified before cloning.

2.3 Cloning of antibody genes from hybridoma RNA

2.3.1 *RNA extraction*

In order to clone immunoglobulin variable genes successfully it is important to obtain high quality ribonucleic acid from the hybridoma cells. Either the hybridoma total RNA or the hybridoma mRNA can be used as a source for cloning the light chain variable region (V_L) and heavy chain variable region (V_H) genes. As starting material for the extraction of RNA, freshly grown productive hybridoma cells should be used. A simple and fast method of extracting RNA suitable for cDNA synthesis from crude cell extracts was published by Ferre and Garduno (1989). The method involves boiling the cells in a solution of diethylpyrocarbonate (DEPC), which inactivates RNAases. Alternatively, a protocol by Sambrook *et al.* (1989) can be used, which relies on the fact that RNAases can be efficiently inactivated in the presence of guanidine isothiocyanate and β-mercaptoethanol. A number of kits for RNA extraction are available commercially, for example, RNA Midi Kit (Qiagen) for the preparation of total RNA, or Dynabeads mRNA Direct Kit (Dynal) for the preparation of mRNA directly from crude cell extracts. cDNA is then prepared from total RNA or mRNA using either oligo(dT) primers, random hexanucleotides or a gene-specific primer and reverse transcriptase (for example, Promega, GIBCO) following the supplier's instructions.

2.3.2 *Primers for the amplification of antibody variable genes*

The Kabat database is a useful source of antibody gene sequences (Johnson *et al.*, 1996). Multiple alignments of immunoglobulin sequences and hypervariability plots have highlighted areas of homology between different antibodies. Areas of sequence identity have been located in the antibody framework. Oligonucleotides can be designed to bind to 5′ and 3′ ends of rearranged V_H and V_L genes in framework areas of sequence homology. Since there are many families of V_L and V_H genes, differing in the amino acid sequence of their framework regions, primers can be designed specifically for each family. Alternatively, degenerate primers can be designed. When cloning repertoires of V genes from spleen cells or peripheral blood lymphocytes from immunized mice (for example, to construct a library of V genes), mixtures of primers can be used to make sure that all of the antibody gene families are amplified (Ørum *et al.*, 1993; Krebber *et al.*, 1997).

Different sets of primers must be used to amplify variable regions with or without parts of the antibody constant region, depending on the cloning format (for example, scFv or Fab), and restriction enzyme sites for cloning are usually incorporated into the primer sequences. For a summary of primers see Table 2.1.

In mice, there are only two V_λ segments, but over 100 V_κ segments. Only 5 per cent of the light chains found in the mouse are V_λ chains (Austyn and Wood, 1993). Some of the existing primer sets might not amplify cDNA from mouse hybridoma cells that secrete V_λ chains (for example, the primer set of the Pharmacia Recombinant Phage Antibody System does not contain λ LC primers; see Pharmacia data file). In some instances, short

Table 2.1 Primers for the cloning of antibody genes

Cloning format	Compatibility with vectors	Primer reference
scFv	pHEN, pCANTAB	Orlandi *et al.* (1989) Ward *et al.* (1989) Clackson *et al.* (1991a,b) Zhou *et al.* (1994) De Jaeger *et al.* (1997) Pharmacia kit
scFv	pAK series	Krebber *et al.* (1997)
Fab	pCOMB3	Kang *et al.* (1991)
Fab		Ørum *et al.* (1993)

stretches of amino acid sequence from an antibody light chain have been used to devise primers (Mason *et al.*, 1996). Monoclonal antibody isotyping kits can be used to establish the immunoglobulin subgroup and the type of light chain (for example IsoDetect, Stratagene; Yang *et al.*, 1998) before cloning the Mab of interest, so that the appropriate primers can be used.

Kits are available commercially that contain primers or primer sets and the vector DNA for the cloning of antibody genes from mouse hybridomas (for example, Pharmacia Recombinant Phage Antibody System Mouse scFv Module; for rationale and protocol see Jackson *et al.*, 1992; Kramer, 1998).

2.3.3 Pseudogenes

Pseudogenes are defective genes that do not give rise to functional protein products. The presence of pseudogene transcripts can create problems during cloning, especially when the functional light or heavy chain mRNAs are transcribed only at a low level. Elevated levels of non-functionally rearranged mRNAs have been found, especially when P3-X63-Ag8.653 SP2/0 myeloma or NS-1 myeloma cells were used as the fusion partner in the production of the parent hybridoma (Strohal *et al.*, 1987; Thammana, 1994). However, for some aberrant mRNAs, it has not been possible to trace their origin back to the myeloma cell lines used for fusions. An alternative explanation suggested by Krebber *et al.* (1997) is that a B-cell may produce aberrant mRNAs in addition to the functional mRNA, or that some hybridoma lines could have arisen by the fusion of several B-cells with one myeloma cell, thus leading to the expression of more than one light and one heavy chain. This was confirmed by Jiang *et al.* (1994) who cloned several different light and heavy chain sequences from a single hybridoma line that produced a Mab which recognizes tick-borne flaviviruses.

A number of methods to circumvent the pseudogene problem have been published. Ostermeier and Michel (1996) suggest the removal of pseudogene RNA before cDNA synthesis by RNAase H digestion of mRNA–DNA hybrids. Such hybrids form when an oligonucleotide homologous to part of a pseudogene is incubated with the mRNA extracted from the hybridoma line. Alternatively, a PCR test can be applied after cloning using a primer specific for the V_{κ}21E pseudogene (present in SP2/0 or P3-X63-Ag8.653) (Strohal *et al.*, 1987; Jovelin *et al.*, 1996). This PCR primer binds to the aberrant light chain, and clones can be screened for its presence.

Phage display (see Chapter 3) of the cloned scFv in 'mini-libraries', to verify functionality by selection against the antigen, is an important approach. Briefly, the antibody is displayed as a fusion protein on the surface of filamentous phage and incubated with the immobilized antigen. After washing, phage that bind to the antigen are eluted and used to infect fresh *E. coli* cells. This results in an enrichment of phage clones that recognize the antigen. Single phage clones can then be tested by ELISA, and only binders are considered for further characterization and the expression of soluble scFv.

2.3.4 Cloning strategies and vectors

Different cloning strategies are employed to assemble light and heavy antibody variable genes as an scFv. An scFv consists of V_H and V_L, joined by a flexible linker. The three-fragment PCR method to assemble V_H, V_L and linker (see Pharmacia kit) can be

..**CGC*GGCCCAGCCGGCC*ATGG**CCCAGGTGCAGCTGCA***GGCGCGCC***TGCA

Sfi I **pelB** *Asc* I

GGTCGACCTCGAGATCAAACGG***GCGGCCGC*****AGAACAAAAACTCATCTC**

Not I **myc tag**

AGAAGAGGATCTGAATGGGGCCGCA***TAG*****ACTGTTGAA**.....

Amber stop **pIII gene**

Figure 2.2 Cloning site of the phagemid vector pCOCK.

problematic, and an alternative procedure that employs sequential cloning was described by Engelhardt *et al.* (1994). Also, two different orientations of the scFv are possible: V_H–linker–V_L (Huston *et al.*, 1988) or V_L–linker–V_H (Bird *et al.*, 1988). Both orientations result in functionally active antibody. It can be helpful first to clone light and heavy chain genes separately into a PCR-cloning vector, then analyse the genes and assemble them in the desired expression vector. This helps to minimize problems arising with the restriction enzyme digestion of PCR products. The restriction sites are usually at the very end of the PCR products, and so more enzyme has to be used and the incubation with the enzyme prolonged to achieve efficient digestion, and subsequently, efficient ligation with the digested vector DNA. We routinely use the vector pGEM-T (Promega) to clone light and heavy chain PCR products. The inserts can then be excised from the pGEM-T vector DNA and assembled in the expression vector of choice.

Vectors available include plasmid vectors, phage vectors and phagemid vectors (Mead and Kemper, 1988). When phage and phagemid vectors are used, the cloned antibody can be displayed on the surface of filamentous phage, for example as a fusion with the minor phage coat protein pIII (Hoogenboom *et al.*, 1991), making selection procedures ('panning') on immobilized antigen possible (see Chapter 3). Phagemid vectors such as pHEN (Hoogenboom *et al.*, 1991) and derivatives contain an amber stop codon (UAG) between the antibody cloning site and pIII (Figure 2.2) which allows an easy switch between soluble scFv production and phage display of the scFv by just using different *E. coli* strains (for example, suppressor strain TG-1 and non-suppressor strain HB2151; Hoogenboom *et al.*, 1991). In the strain HB2151 the protein terminates at the amber (UAG) codon, resulting in soluble scFv protein. In the suppressor strain TG-1, glutamic acid will be incorporated in the protein molecule and the scFv will be fused to the pIII phage coat protein.

Engelhardt *et al.* (1994) have published a modified version of the pHEN vector (Hoogenboom *et al.*, 1991) that allows the sequential cloning of V_H and V_L. To construct the pCOCK vector, an *Asc* I site was inserted between the *Sfi* I and the *Not* I sites of the pHEN vector (Figure 2.2). For the V_H primers, recognition sites for *Sfi* I and *Asc* I were incorporated; for the V_L primers, recognition sites for *Asc* I and *Not* I. The V region specific primers designed for use with this vector contain the sequence for the linker. Thus, heavy and light chains can be cloned separately and sequentially, overcoming the need to assemble the full-length scFv construct by PCR. The abovementioned restriction enzymes were chosen because they were identified as being rare cutters in antibody sequences.

It is therefore highly unlikely that antibody sequences will be inactivated by restriction during the cloning process.

The recognition sequence for *Sfi* I is 13 nucleotides long; the central five nucleotides can consist of any combination of nucleotides. This property of *Sfi* I is the basis of a recently published directional cloning system that utilizes only one enzyme (*Sfi* I) and two different sticky ends (Krebber *et al.*, 1997).

Clones can be digested with restriction enzymes such as *BstN* I (Marks *et al.*, 1991; see Chapter 3 for protocol) or *Mva* I (De Jaeger *et al.*, 1997). This, in combination with sequencing of a number of clones, should show up sequence heterogeneity due to pseudogenes or mutations.

Antibody sequences can be analysed for their antibody group/family by comparing them against known antibody sequences using DNAPLOT (http://www.genetik.uni-koeln.de) or the Kabat database (http://immuno.bme.nwu.edu; Johnson *et al.*, 1996).

2.4 Expression of scFv in *E. coli*

After the initial successful cloning of the antibody genes from the hybridoma line, problems can arise with large scale expression of the soluble antibodies, for example the antibody sequence may prove to be toxic for *E. coli* cells, or insoluble antibody may aggregate in the *E. coli* cells. Different strains of *E. coli* have been used for the expression of antibody genes in attempts to optimize yield (Dueñas *et al.*, 1994; Ayala *et al.*, 1995). Expression levels in a given strain and the localization of the scFv (for example, periplasmic space or culture supernatant) vary with different antibody fragments and are sequence dependent. These problems are discussed in detail in Chapter 4.

Clones that contain the assembled light and heavy chains should be tested for scFv expression under different growth conditions to determine the influence of temperature, induction times and culture media on yield and antibody function. The recombinant antibody protein can accumulate in the cytoplasm, the periplasmic space or even leak out into the culture medium, depending on the antibody sequence (Ayala *et al.*, 1995). Therefore, the expression conditions have to be optimized for each new scFv. The presence of positive charges (amino acids arginine and lysine) within 20–30 amino acids downstream of the signal peptide is thought to inhibit protein translocation to the periplasm (Ayala *et al.*, 1995).

Point mutations can have a big influence (Dueñas *et al.*, 1995) on the yield and localization of recombinant antibodies. For example, a mutant anti-cucumber mosaic virus scFv II/6 with two changes in the V_L CDR3 showed an approximately tenfold increase in expression compared with the parent scFv (unpublished results). Forsberg *et al.* (1997) have identified framework residues in a secreted recombinant antibody fragment which control expression level and localization in *E. coli* (for example, periplasmic space or culture supernatant). Mutations in the V_L mainly seemed to influence yield, whereas mutations in the V_H influenced the final product localization.

2.4.1 *Purification of scFv*

Depending on the localization of the recombinant protein and the tags used, different strategies should be employed to harvest and purify the product (for protocols see Chapter 4). Proteases can affect yield and antibody performance, depending on the amino acid composition of each individual antibody sequence. Therefore, failure to detect the expression of a recombinant antibody in *E. coli* may simply be a consequence of rapid proteolysis, and protease inhibitors may help to prevent degradation.

The desired level of purity for a recombinant antibody depends on the ultimate application of the product. For therapeutic applications, high purity is of course needed. Contaminating proteins, bacterial endotoxins, bacterial nucleic acids and bacterial viruses have to be removed in order to avoid complications in patients treated with the recombinant product. On the other hand, it may not be necessary to purify recombinant antibody for routine ELISA applications; particularly if the recombinant antibody is stable in a crude bacterial extract, it can be used as detecting reagent for plant pathogens.

The antibodies can be concentrated by ammonium sulphate precipitation from the culture supernatant or harvested as a periplasmic extract from a cell pellet, dialysed against phosphate-buffered saline (PBS) and used directly, or stored until needed. scFv are more sensitive than Fab fragments or whole antibody to destabilization by different agents. For example, glycerol seems to promote the dissociation of the two variable domains of the scFv protein, but does not affect the domain structure of scFv (Muller *et al.*, 1994). This has to be taken into account when deciding how to store the scFv.

We have successfully freeze-dried our antibodies, stored, reconstituted and then used them in ELISA and other tests. After freeze-drying, the antibodies can be sent at ambient temperature by ordinary post to testing laboratories. The antibodies remained functional for at least 12 months after freeze-drying.

If the recombinant antibody is to be used as a coating reagent for ELISA plates in order to trap antigen from test samples (for example, plant extracts), a higher degree of purity is needed. *E. coli* proteins present in the crude preparation could also bind to the ELISA plate, thereby reducing the amount of recombinant antibody binding to the plate. This could lead to less efficient trapping of the antigen from plant sap. We have successfully used a recombinant antibody, containing a human κ light chain constant region, for trapping of *Tomato yellow leafcurl virus* (genus *Begomovirus*) from leaf extracts of infected tomato plants. The antibody preparation used was purified by immobilized metal affinity chromatography (IMAC; Hochuli *et al.*, 1988; Lindner *et al.*, 1992; for protocol, see Chapter 4).

2.4.2 Clone maintenance

For clone maintenance it is advisable to keep either frozen bacterial stocks at –70°C or purified plasmid DNA. It is not recommended to keep recombinant bacteria on agar plates for long periods of time, because this could lead to the accumulation of mutations, deletion of the scFv insert and ultimately to the loss of antigen binding. All incubations of plates and cultures should be done at ≤ 30°C (see also Chapter 4 for influence of culture temperature on insert stability).

2.5 Affinity of scFv

Often the apparent affinity of the recombinant antibody does not compare favourably with the affinity of the Mab secreted by the parent hybridoma. There are different reasons for this phenomenon. Firstly, when the recombinant antibody is cloned in the scFv format, it will only exhibit monovalent binding to the antigen (as opposed to bivalent binding by the parent antibody). Secondly, the arrangement of light and heavy chain via a short linker in the scFv might not always be sufficiently conformationally stable to recreate the exact binding pocket of the parent antibody. It has been reported that scFv can be more cross-reactive than the parent antibody (Yuan *et al.*, 1997). One reason for this could be a slight conformational distortion (small differences in the relative packing of the V_H and V_L domains) of the scFv binding pocket because of the loss of constraint normally imposed by the constant region (Whitlow *et al.*, 1994; Zdanov *et al.*, 1994).

When purified preparations of scFv were analysed, it was noted that monomers and dimers were present (Griffiths *et al.*, 1993), and that the dimers exhibited bivalent binding to the antigen. This has led to the idea of designing so-called diabodies (Holliger *et al.*, 1993). If the linker is short enough, the V_H and the V_L domains have to pair with the complementary domains of another scFv molecule to form a dimer. If there is no linker at all, the formation of trimers is possible, but it has not been established how these would perform as detecting reagents in immunoassays (Kortt *et al.*, 1997). Diabodies can enhance binding of the recombinant antibody to its target by making it bivalent. The combination of two scFv with different binding specificities in the diabody format allows simultaneous binding to two different targets (bispecific diabodies, for example, intended for tumour-targeting) (Hoogenboom, 1997). The diabody format makes it possible to combine the antigen recognition with a biological effect.

Functional bivalency can also be restored by adding a light chain constant region to the scFv construct (for example, pPM1 (McGregor *et al.*, 1994) or pZip and pC_LZip (Kerschbaumer *et al.*, 1997; see also Chapter 4)). scFv produced from these constructs tend to dimerize spontaneously. Dimerization also takes place when the antibody is expressed as a fusion protein with *E. coli* alkaline phosphatase, because the active form of the enzyme is a dimer.

The binding affinity for both monoclonal and recombinant antibodies can be estimated using a competition ELISA (Rath *et al.*, 1988). The phage ELISA described by Dyson *et al.* (1995) is a variation of the competition assay, which gives an estimate of the affinity of scFv still displayed on the surface of phage. Other methods to measure binding affinities, including surface plasmon resonance, are described in detail in Chapter 5.

2.6 Application of scFv to plant virus diagnosis

We have cloned and expressed scFv genes from two hybridoma lines secreting monoclonal antibodies to *African cassava mosaic virus* (ACMV, genus *Begomovirus*). Mab SCR 20 and Mab SCR 23 have been produced against ACMV, a whitefly-transmitted geminivirus, but they cross-react with other whitefly-transmitted viruses, which cause yellow leaf curl diseases in tomato (*Tomato yellow leaf curl virus*, TYLCV, genus *Begomovirus*; Thomas *et al.*, 1986).

Mabs SCR 20 and 23 were cloned as scFv in the V_H–linker–V_L orientation in the vector pCOCK as described in this chapter. To study the influence of the orientation, a V_L–linker–V_H version of the SCR 20 Mab was cloned in pCOCK. For SCR 20, the orientation of the scFv fusion did not affect the ELISA results. Both SCR 20 and SCR 23 were also cloned in the pPM1-His vector as fusions to a human κ constant region.

2.6.1 *ELISA*

ScFv expressed from pCOCK carry the myc tag and represent mostly monovalent binding. In the pPM1-His vector, a human κ constant region is fused to the scFv. The fusion

Table 2.2 Comparison of Mab SCR 20, scFv 20 and scFvK 20 in ELISA

Virus strain	Mab20	scFv20	scFvK20	scFvK20*
ACMV	>1.8**	0.95	>1.8	>1.8
TYLCV-Sardinia	0.3	0.25	0.4	0.4
Healthy tomato	0.06	0.07	0.07	0.07

* Recombinant antibody was freeze-dried and reconstituted before use in ELISA.
** A_{405nm} measured after 1h incubation with substrate buffer.

proteins (scFvK) dimerize spontaneously, leading to bivalent binding (McGregor *et al.*, 1994). Both types of recombinant antibodies can detect ACMV and TYLCV strains in a triple antibody sandwich (TAS) ELISA. In this type of ELISA, plant sap is added to the wells of ELISA plates coated with a polyclonal antiserum prepared against ACMV. The recombinant antibodies are then used as the detecting agent. Binding of the recombinant antibodies was shown either using an anti-c-myc tag antibody (Evan *et al.*, 1985) or an anti-human κ constant region alkaline phosphatase conjugate (Sigma) (Table 2.2; see also Torrance *et al.*, 1998a, 1998b). Parent Mab and scFv showed comparable results.

The purified scFvK was also used to coat microtitre plates to capture the virus successfully from plant extract; the detecting agent in this instance was the parental monoclonal antibody.

2.6.2 *Tissue printing*

From earlier work it was known that both SCR20 and SCR23 bind to discontinuous epitopes on the surface of the virus coat protein. The Pepscan technique (Geysen *et al.*, 1987) did not identify peptides recognized by SCR 20. Also, the Mabs did not recognize denatured viral coat protein on Western blots (Swanson, 1992).

However, it was shown that the Mabs can be used to detect virus particles in tissue prints (Fargette *et al.*, 1996). Therefore, the recombinant antibodies were tested for their ability to detect virus particles in tissue prints of stems of TYLCV-infected *L. esculentum* and were found to perform well in comparison with the parental Mabs (Torrance *et al.*, 1998a, 1998b). The recombinant antibodies could be used to study the localization and distribution of virus particles in infected tissue. The virus particles were visualized in the vascular bundles as purple spots due to the alkaline phosphatase substrate used. For use in tissue printing the recombinant antibodies did not have to be purified. Concentrated culture supernatant, containing recombinant antibody was sufficient and did not result in a high background.

2.7 Conclusions

The genes encoding the variable regions of antibodies can be cloned from RNA prepared from hybridoma cells. This method is especially valuable for the rescue of unstable hybridomas. It allows the cloning of valuable, well characterized Mabs and the use of their variable antibody genes for biotechnology applications.

The antibody DNA can be stored in the freezer, and it can be genetically manipulated to produce immunological reagents of choice (see Chapter 4). Utilizing suitable expression vectors, reliable recombinant reagents for use in immunoassays to detect plant viruses can be obtained from *E. coli* cultures. These recombinant reagents can be produced quickly and cheaply without the use of animals or large-scale cell culture, and have the potential to complement or replace polyclonal and monoclonal antibodies in immunoassays.

The sequences obtained from hybridomas by molecular cloning, just like the sequences selected from phage display libraries, are not limited to applications in diagnostics. They can also be expressed in plants to engineer resistance to pathogens (Chapter 10) or pests (Chapter 11), and to alter metabolic pathways (Chapter 9).

Acknowledgements

The Scottish Office Agriculture, Environment and Fisheries Department and the European Union (AIR CT 96–1046) are gratefully acknowledged for funding. I also thank G. Himmler and W. Harrison for providing cloning vectors.

References

AUSTYN, J.M., WOOD, K.J., 1993, *Principles of Cellular and Molecular Immunology*, Oxford University Press.

AYALA, M., BALINT, R.F., FERNANDEZ DE COSSIO, M.E., CANAAN-HADEN, L., LARRICK, J.W. and GAVILONDO, J.V., 1995, Variable region sequence modulates periplasmic export of a singlechain Fv antibody fragment in *Escherichia coli*, *BioTechniques*, **18**, 832–842.

BETTER, M., CHANG, C.P., ROBINSON, R.R. and HORWITZ, A.H., 1988, *Escherichia coli* secretion of an active chimeric antibody fragment, *Science*, **240**, 1041–1043.

BIRD, R.E., HARDMAN, K.D., JACOBSON, J.W., JOHNSON, S., KAUFMANN, B.M., LEE, S.M., *et al.*, 1988, Single-chain antigen-binding proteins, *Science*, **242**, 423–426.

CLACKSON, T., GÜSSOW, D. and JONES, P.T., 1991a, General applications of PCR to gene cloning and manipulation, in *PCR: A Practical Approach*, edited by M.J. MCPHERSON, P. QUIRKE and G.R. TAYLOR, IRL Press.

CLACKSON, T., HOOGENBOOM, H.R., GRIFFITHS, A.D. and WINTER, G., 1991b, Making antibody fragments using phage display libraries, *Nature*, **352**, 624–628.

DE JAEGER, G., BUYNS, E., EECKHOUT, D., BRUYNS, A.-M., DE NEVE, M., DE WILDE, C., *et al.*, 1997, Use of phage display for isolation and characterization of single-chain variable fragments against dihydroflavonol 4-reductase from *Petunia hybrida*, *FEBS Letters*, **403**, 116–122.

DUEÑAS, M., VAZQUEZ, J., AYALA, M., SÖDERLIND, E., OHLIN, M., PEREZ, L., *et al.*, 1994, Intra- and extracellular expression of an scFv antibody-fragment in *E. coli*: effect of bacterial strains and pathway engineering using GroES/L chaperonins, *BioTechniques*, **16**, 476–486.

DUEÑAS, M., AYALA, M., VAZQUEZ, J., OHLIN, M., SÖDERLIND, E., BÖRREBAECK, C.A.K. and GAVILONDO, J.V., 1995, A point mutation in a murine immunoglobulin V-region strongly influences the antibody yield in *E. coli*, *Gene*, **158**, 61–66.

DYSON, M.R., GERMASCHEWSKI, V. and MURRAY, K., 1995, Direct measurement via phage titre of the dissociation constants in solution of fusion phage-substrate complexes, *Nucleic Acids Research*, **23**, 1531–1535.

ENGELHARDT, O., GRABHERR, R., HIMMLER, G. and RÜKER, F., 1994, Two-step cloning of antibody variable domains in a phage-display vector, *BioTechniques*, **17**, 44–46.

EVAN, G.I., LEWIS, G.K., RAMSEY, G. and BISHOP, J.M., 1985, Isolation of monoclonal antibodies specific for human c-myc proto-oncogene product, *Molecular and Cellular Biology*, **5**, 3610–3616.

FARGETTE, D., LESLIE, M. and HARRISON, B.D., 1996, Serological studies on the accumulation and localisation of three tomato leaf curl geminiviruses in resistant and susceptible Lycopersicon species and tomato cultivars, *Annals of Applied Biology*, **128**, 317–328.

FERRE, F. and GARDUNO, F., 1989, Preparation of crude cell extract for amplification of RNA by the polymerase chain reaction, *Nucleic Acids Research*, **17**, 2141.

FORSBERG, G., FORSGREN, M., JAKI, M., NORIN, M., STERKY, C., ENHÖRNING, A., *et al.*, 1997, Identification of framework residues in secreted recombinant antibody fragments that control production level and localization in *Escherichia coli*, *Journal of Biological Chemistry*, **272**, 12430–12436.

GEYSEN, H.M., RODDA, S.J., MASON, T.J., TRIBBICK, G. and SCHOOFS, P.G., 1987, Strategies for epitope analysis using peptide synthesis, *Journal of Immunological Methods*, **102**, 259–274.

GRIFFITHS, A.D., MALMQUIST, M., MARKS, J.D., BYE, J.M., EMBLETON, M.J., MCCAFFERTY, J., *et al.*, 1993, Human anti-self antibodies with high specificity from phage display libraries, *EMBO Journal*, **12**, 725–734.

HOCHULI, E., BANNWARTH, W., DÖBERLI, H., GENTZ, R. and STÜBER, D., 1988, Genetic approach to facilitate purification of recombinant proteins with a novel metal chelate absorbent, *Biotechnology*, **6**, 1321–1325.

HOLLIGER, P., PROSPERO, T. and WINTER, G., 1993, 'Diabodies': small bivalent and bispecific antibody fragments, *Proceedings of the National Academy of Science USA*, **90**, 6444–6448.

HOOGENBOOM, H.R., 1997, Mix and match: building manifold binding sites, *Nature Biotechnology*, **15**, 125–126.

HOOGENBOOM, H.R., GRIFFITHS, A.D., JOHNSON, K.S., CHISWELL, D.J., HUDSON, P. and WINTER, G., 1991, Multi-subunit proteins on the surface of filamentous phage: methodologies for displaying antibody (Fab) heavy and light chains, *Nucleic Acids Research*, **19**, 4133–4137.

HUSTON, J.S., LEVINSON, D., MUDGETT-HUNTER, M., TAI, M.S., NOVOTNY, J., MARGOLIES, M.N., *et al.*, 1988, Protein engineering of antibody binding sites: recovery of specific activity in an anti-digoxin single-chain Fv analogue produced in *Escherichia coli*, *Proceedings of the National Academy of Science USA*, **85**, 5879–5883.

JACKSON, R.H., MCCAFFERTY, J., JOHNSON, K.S., POPE, A.R., ROBERTS, A.J., CHISWELL, D.J., *et al.*, 1992, Selection of variants of antibodies and other protein molecules using display on the surface of bacteriophage fd, in *Protein Engineering: A Practical Approach*, Rees, A.R., STERNBERG, M.J.E. and WETZEL, R., Eds, pp. 277–301.

JIANG, W., BONNERT, T.P., VENUGOPAL, K. and GOULD, E., 1994, A single chain antibody fragment expressed in bacteria neutralizes tick-borne flaviviruses, *Virology*, **200**, 21–28.

JOHNSON, G., KABAT, E.A. and WU, T.T., 1996, Kabat database of sequences of proteins of immunological interest, in *Weir's Handbook of Experimental Immunology I. Immunochemistry and Molecular Immunology*, fifth edition, Eds L.A. HERZENBERG, W.M. WEIR and C. BLACKWELL, Blackwell Science Inc., Cambridge, MA, Chapter 6, pp. 6.1–6.21.

JOVELIN, F., BRARD, F., RAUX, G., TRON, F. and GILBERT, D., 1996, Immunoglobulin variable-region mRNA direct sequencing: a method to bypass aberrant myeloma light-chain transcripts, *BioTechniques*, **21**, 576–580.

KANG, A.S., BURTON, D.R. and LERNER, R.A., 1991, Combinatorial immunoglobulin libraries in phage λ, *Methods: A companion to Methods in Enzymology*, **2**, 111–118.

KERSCHBAUMER, R.J., HIRSCHL, S., KAUFMANN, A., IBL, M., KOENIG, R. and HIMMLER, G., 1997, Single-chain Fv fusion proteins suitable as coating and detecting reagents in a double antibody sandwich enzyme linked immunosorbent assay, *Analytical Biochemistry*, **249**, 219–227.

KÖHLER, G. and MILSTEIN, C., 1975, Continuous cultures of fused cells secreting antibody of predefined specificity, *Nature*, **256**, 495–497.

KORTT, A.A., LAH, M., ODDIES, G.W., GRUEN, C.L., BURNS, J.E., PEARCE, *et al.*, 1997, Single-chain Fv fragments of anti-neuraminidase antibody NC10 containing five- and ten-residue linkers form dimers and with zero-residue linker a trimer, *Protein Engineering*, **10**, 423–433.

KRAMER, K., 1998, Synthesis of pesticide-specific single-chain Fv utilizing the recombinant phage antibody system (RPAS, Pharmacia). A detailed protocol, *Analytical Letters*, **31**, 67–92.

KREBBER, A., BORNHAUSER, S., BURMESTER, J., HONEGGER, A., WILLUDA, J., BOSSHARD, H.R. and PLÜCKTHUN, A., 1997, Reliable cloning of functional antibody variable domains from hybridomas and spleen cell repertoires employing a reengineered phage display system, *Journal of Immunological Methods*, **201**, 35–55.

LINDNER, P., GUTH, B., WÜLFING, C., KREBBER, C., STEIPE, B., MÜLLER, F. and PLÜCKTHUN, A., 1992, Purification of native proteins from cytoplasm and periplasm of *Escherichia coli* using IMAC and histidine tails: a comparison of proteins and protocols, *Methods: A Companion to Methods in Enzymology*, **4**, 41–56.

MARKS, J.D., HOOGENBOOM, H.R., BONNERT, T.P., MCCAFFERTY, J., GRIFFITHS, A.D. and WINTER, G., 1991, By-passing immunization: human antibodies from V-gene libraries displayed on phage, *Journal of Molecular Biology*, **222**, 581–597.

MASON, P., BERNSTEIN, A., BAXT, B., PARSELLS, R., KANG, A. and RIEDER, E., 1996, Cloning and expression of a single-chain antibody fragment specific for foot-and-mouth disease virus, *Virology*, **224**, 548–554.

MCGREGOR, D.P., MOLLOY, P.E., CUNNINGHAM, C. and HARRIS, W.J., 1994, Spontaneous assembly of bivalent single chain antibody fragments in *Escherichia coli*, *Molecular Immunology*, **31**, 219–226.

MEAD, D.A. and KEMPER, B., 1988, Chimeric single-stranded DNA phage-plasmid cloning vectors, in *Vectors: A Survey of Molecular Cloning Vectors and their Uses*, Eds R.L. RODRIQUES and D.T. DENHARDT, Bostob, Butterworths, pp. 85–102.

MULLER, J.D., NIENHAUS, G.U., TETIN, S.Y. and VOSS, E.W., 1994, Ligand-binding to anti-fluorescyl antibodies stability of the antigen-binding site, *Biochemistry*, **33**, 6221–6227.

MULLIS, K., FALOONA, F., SCHARF, S., SAIKI, R. and EHRLICH, H., 1986, Specific enzymatic amplification of DNA *in vitro*: the polymerase chain reaction, *Cold Spring Harbor Symposium on Quantitative Biology*, **51**, 263–273.

MUNRO, S. and PELHAM, H.R.B., 1986, An Hsp-like protein in the ER: identity with the 78 kD glucose regulated protein and immunoglobulin heavy chain binding protein, *Cell*, **46**, 291–300.

ORLANDI, R., GÜSSOW, D.H., JONES, P.T. and WINTER, G., 1989, Cloning immunoglobulin variable domains for expression by the polymerase chain reaction, *Proceedings of the National Academy of Science USA*, **86**, 3833–3837.

ØRUM, H., ANDERSEN, P.S., OSTER, A., JOHANSEN, L.K., RIISE, E., BJORNVAD, M., *et al.*, 1993, Efficient method for constructing comprehensive murine Fab antibody libraries displayed on phage, *Nucleic Acids Research*, **21**, 4491–4498.

OSTERMEIER, C. and MICHEL, H., 1996, Improved cloning of antibody variable regions from hybridomas by an antisense-directed RNAse H digestion of the P3-X63-Ag8.653 derived pseudogene mRNA, *Nucleic Acids Research*, **24**, 1979–1980.

RATH, S., STANLEY, C.M. and STEWARD, M.W., 1988, An inhibition enzyme immunoassay for estimating relative antibody affinity and affinity heterogeneity, *Journal of Immunological Methods*, **106,** 245–249.

SAMBROOK, J., FRITSCH, E.F. and MANIATIS, T., 1989, *Molecular Cloning: A Laboratory Manual*, 2nd edition, Cold Spring Harbor Laboratory Press, Cold Spring Harbor, NY, Vol. 1–3.

STROHAL, R., KROEMER, G., WICK, G. and KOFLER, R., 1987, Complete variable-region sequence of a nonfunctionally rearranged kappa light chain transcribed in the nonsecretor P3-X63-Ag8.653 myeloma cell line, *Nucleic Acids Research*, **15**, 2771.

SWANSON, M.M., 1992, Antigenic variation and serological relationship of whitefly-transmitted geminiviruses, University of Dundee, PhD Thesis.

TERRADA, E., KERSCHBAUMER, R., HIMMLER, G., GIUNTA, G., GALEFFI, P., GORRIS, M.T. and CAMBRA, M., 1998, Two engineered antibodies (3DF1 and 3CA5 scFv) specific for citrus tristeza virus reacting in ELISA as conventional monoclonal antibodies, 7th International

Congress of Plant Pathology, Edinburgh, Scotland, 1998, *Offered Papers Abstracts*, Vol. 3, 3.3.30.

THAMMANA, P., 1994, Characterization of a deletion in the P3 myeloma heavy chain V region sequence in the non-secreting cell line NS-1, *Molecular Immunology*, **31**, 77–78.

THOMAS, J.E., MASSALSKI, P.R. and HARRISON, B.D., 1986, Production of monoclonal antibodies to African cassava mosaic virus and differences in their reactivities with other whitefly-transmitted geminiviruses, *Journal of General Virology*, **67**, 2739–2748.

TORRANCE, L., ZIEGLER, A., HARPER, K. and MAYO, M.A., 1998a, Engineered antibodies: readily adaptable molecular tools for basic and applied research, Scottish Crop Research Institute, Annual Report 1997/98, 111–113.

TORRANCE, L., ZIEGLER, A., HARPER, K., MACINTOSH, S., ZANGE, B., KUMARI, S. and MAYO, M., 1998b, Novel molecular tools for plant virus detection: recombinant antibodies specific for luteoviruses and geminiviruses, *Proceedings of VII International Congress for Plant Pathology*, 9–16 August, Edinburgh.

WARD, E.S., GÜSSOW, D.H., GRIFFITHS, A.D., JONES, P.T. and WINTER, G., 1989, Binding activities of a repertoire of single immunoglobulin variable domains secreted from *E. coli*, *Nature*, **341**, 544–546.

WHITLOW, M., FILPULA, D., ROLLENCE, M.L., FENG, S-L. and WOOD, J.F., 1994, Multivalent Fvs: characterization of single-chain Fv oligomers and preparation of a bispecific Fv, *Protein Engineering*, **7**, 1017–1026.

YANG, W., PETRE, M.C., HOLMES, D. and PINGERELLI, P., 1998, Simple, sensitive isotyping of mouse monoclonal antibodies, *Strategies*, **10**, 75–76.

YUAN, Q., CLARKE, J.R., ZHOU, H.-R., LINZ, J.E., PESTKA, J.J. and HART, P., 1997, Molecular cloning, expression and characterization of a functional single-chain Fv antibody to the mycotoxin zearalenone, *Applied and Environmental Microbiology*, **63**, 263–269.

ZDANOV, A., LI, Y., BUNDLE, D.R., DENG, S.J., MACKENZIE, C.R., NARANG, S. and CYGLER, M., 1994, Structure of a single-chain antibody variable domain (FV) complexed with a carbohydrate antigen at 1.7 angstroem resolution, *Proceedings of the National Academy of Science USA*, **91**, 6423–6427.

ZHOU, H., FISHER, R.J. and PAPAS, T.S., 1994, Optimization of primer sequences for mouse scFv repertoire display library construction, *Nucleic Acids Research*, **22**, 888–889.

3

Selection of Anti-Plant Virus Single-Chain Variable Fragments from Phage Display Antibody Libraries

KAREN HARPER, RACHEL L. TOTH, MIKE A. MAYO AND LESLEY TORRANCE

3.1 Introduction

Antibody-based tests have been used extensively for the detection of a wide range of plant pathogens, including viruses (Torrance, 1992), phytoplasmas (Clark, 1992), fungi and bacteria (Dewey, 1992; Stead, 1992). In particular, monoclonal antibodies secreted *in vitro* by cultured hybridoma cells have proved to be useful sources of specific diagnostic reagents for plant diseases (Torrance, 1995). However, the procedures for their production and maintenance have some disadvantages. These include the need for specialized mammalian cell culture facilities and the costly storage of cell lines in liquid nitrogen.

The development of phage display technology and the production of recombinant antibodies in *Escherichia coli* was pioneered in the medical sciences to produce antibodies suitable for human therapy. Since then, the technology has been adopted and modified for use in other areas of research, including the plant sciences. Recombinant antibodies can be produced from antibody libraries quickly and cheaply without the need for specialized equipment, and often without the need to immunize animals. Such antibodies have the potential to provide unlimited supplies of specific reagents that have binding properties similar to those of polyclonal and monoclonal antibodies.

This chapter introduces 'phage display' and describes some methods for selecting antigen-specific antibody fragments (single-chain variable fragments (scFv) and Fab fragments) from antibody libraries. In particular, we describe the selection of scFv to particles of *Potato leafroll virus* (PLRV; genus *Polerovirus*) and how these have been adapted for use in immunoassays to detect PLRV in potato tissue. Detailed protocols are included which, combined with those described in Chapter 4, provide comprehensive methods for the production of scFv from a phage display library.

3.2 Production of phage display libraries

The advent of the polymerase chain reaction (PCR) and the increased knowledge of antibody gene sequences (Johnson *et al.*, 1996; Tomlinson *et al.*, 1996) has made it possible to produce large repertoires of antibody fragments. Oligonucleotide sequences have been designed that bind to the highly conserved terminal regions of the V_H and V_L genes of different animal species (see Chapter 2). By using sets of oligonucleotide primers, the antibody genes from mice (Clackson *et al.*, 1991; Hoogenboom *et al.*, 1991), human donors (Vaughan *et al.*, 1996; Sheets *et al.*, 1998), rabbits (Ridder *et al.*, 1995) and camels (Hamers-Casterman *et al.*, 1993) have been amplified by PCR. The products were then assembled in plasmid vectors to produce repertoires of antibody fragments.

Antibody fragments are expressed in *E. coli* most often in the form of scFv. The polypeptides that correspond to the antibody V_H and V_L regions are usually joined with a short linker sequence, typically of 15–20 amino acids, and fused to the N-terminus of the phage pIII minor coat protein. The resultant scFv–pIII fusion product is expressed on the surface of the mature phage particles (McCafferty *et al.*, 1990). Fab fragments have also been produced in which one chain is fused to pIII and the other is secreted into the bacterial periplasm (Barbas III *et al.*, 1991; Hoogenboom *et al.*, 1991). Also, Fab fragments have been expressed fused to the pVIII coat protein (Kang *et al.*, 1991b).

3.2.1 Antibody libraries

The production of large libraries of scFv, Fab fragments, or modified versions of these (such as diabodies) have been reported (for example, Barbas III *et al.*, 1991; Hoogenboom *et al.*, 1991; Williamson *et al.*, 1993; Griffiths *et al.*, 1994; Nissim *et al.*, 1994; McGuinness *et al.*, 1996; Vaughan *et al.*, 1996; Sheets *et al.*, 1998). Detailed methods for the production of phage display libraries have been reported (Pope *et al.*, 1996; Kramer, 1998) and the required reagents (primers, vectors, etc.) are commercially available (Pharmacia Biotech).

Sources of immunoglobulin genes include the spleens and peripheral blood lymphocytes of immunized animals (for example, Clackson *et al.*, 1991) or the bone marrow, tonsils and peripheral blood lymphocytes from non-immunized human donors (Vaughan *et al.*, 1996; Sheets *et al.*, 1998). Libraries produced from immunized donors contain representatives of antibody genes after their rearrangement and after somatic mutation of the germline sequences *in vivo*. These repertoires are biased towards antibody sequences that bind to the immunogen, whereas libraries constructed from non-immunized donors contain populations of antibody sequences that correspond to those of the primary repertoire as well as memory B-cells (the antibody sequences after somatic mutation of the germline cells) of the individuals from which the samples were taken. The bias of the scFv sequences in these libraries is dependent on the medical history of the human donors, but they are not intentionally biased towards any particular antigen.

In contrast, ‘synthetic’ libraries have been made by cloning germline genes and combining these with synthetic CDR3 sequences. Nissim *et al.* (1994) produced a scFv library with a diversity of $>10^8$ clones by incorporating a mixture of synthetic V_H CDR3 regions (of between four and 12 amino acids) produced by PCR into 49 cloned human germline V_H genes. The PCR products were ligated to a single $V_{\lambda 3}$ light chain segment that also incorporated a short linker sequence, and then cloned into a phagemid display vector. The cloning procedures were designed to mimic *in vitro* the process of variable domain rearrangements (V_H–D_H–J_H gene assembly) responsible for generating antibody diversity *in vivo*.

An ‘immunized’ library has the disadvantages that a new library is needed for each antigen and selection is dependent on the immune response to the immunogen, whereas ‘non-immunized’ or ‘synthetic’ libraries can be used as a source of antibody fragments for many antigens (for example, Griffiths *et al.*, 1994; Nissim *et al.*, 1994; Vaughan *et al.*, 1996; Sheets *et al.*, 1998).

It is thought that as the size of the repertoire of antibody fragments in a library prepared from immunized donor material increases, the greater becomes the likelihood of there being phage antibodies in the library that consist of the original pairs of V_H and V_L sequences expressed by the original B-cells *in vivo*. Therefore, it is more likely that high affinity binders may be selected from such large libraries (Winter and Milstein, 1991; Gherardi and Milstein, 1992). Similarly, for libraries derived from non-immunized human donors, scFv selected from libraries with 10^7–10^8 members often have binding affinities in the micromolar range, and are thus characteristic of antibodies of a primary immune response (Marks *et al.*, 1991; Griffiths *et al.*, 1993). scFv with subnanomolar affinities (characteristic of antibodies of a secondary immune response) have been selected from libraries with between 10^9 and 10^{10} members (Vaughan *et al.*, 1996; Sheets *et al.*, 1998).

Although larger libraries contain scFv genes with a potentially greater number of different sequences, some of these may contain internal stop codons, or the expressed protein is not expressed efficiently, or folded correctly in bacterial cells. To try to overcome these disadvantages, fully synthetic libraries have been constructed. In these, framework V_H and V_L sequences have been combined with synthetic CDR sequences that comprise mixtures of trinucleotide cassettes, which are biased against stop codons and which contain codons favoured by bacterial protein translation systems. It is claimed that this approach results in the display of a greater proportion of functional scFv (Krebs *et al.*, 1998).

3.2.2 General principles of selecting antibodies from libraries

The probability of picking, at random, an individual phage antibody with a high affinity to an antigen of interest, from the very large number of clones within a library is extremely small. Therefore, it is necessary to enrich for antigen-binding clones in the phage preparation before testing individual phage clones for antigen-binding molecules. Several different methods have been devised to select specific, individual phage antibody clones from very large antibody libraries.

Enriching the phage preparation for specific antigen-binding antibody fragments is done by a series of selection steps (panning) (see Figure 3.1 and Section 3.5). Briefly, a phage preparation (representing approximately 10^{12} colony forming units, prepared by expanding the library) is incubated with the antigen. Phage that carry antibodies that bind to antigen are separated from unbound phage and the bound phage are then eluted from the antigen and used to infect *E. coli* cells. Phage are harvested from the infected bacterial cultures and used for further selection. It has been shown that by one round of panning, specific phage can be enriched by 20 to 1000 times, or more (Marks *et al.*, 1992a). Thus, only a few rounds of panning are necessary to identify individual phage expressing scFv with the required binding characteristics.

Panning is often done using antigen immobilized on polystyrene tubes (Figure 3.1). If a suitable assay is available it is advisable to confirm the amount of antigen immobilized. It is sometimes necessary to conjugate low molecular weight molecules to larger carrier proteins to ensure adequate immobilization of the antigen for presentation to the library (Harrison *et al.*, 1996; Johnson and Hawkins, 1996). Alternatively, the panning may be done with the antigen in solution. For example, biotin-labelling of the antigen allows biotinylated antigen–phage complexes to be removed from unbound phage by binding to streptavidin-coated magnetic beads (Figure 3.1). Also, biotin-labelling of antigen enables

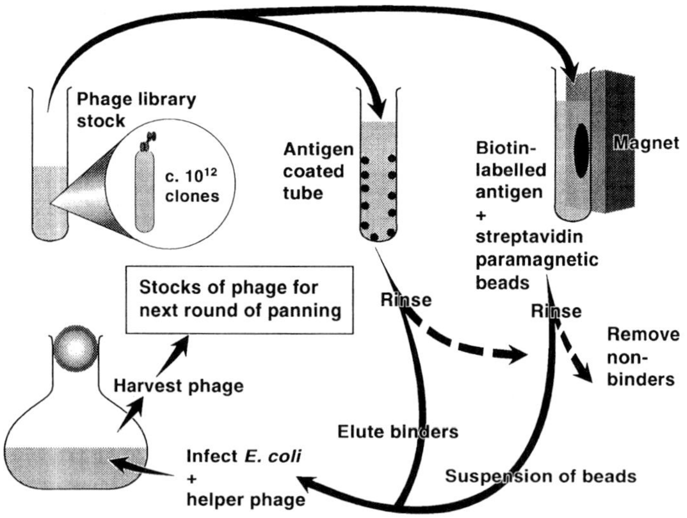

Figure 3.1 Illustration of the procedure for one round of selection (panning) against immobilized antigen, or biotinylated antigen in solution, from a phage display library.

successively lower concentrations of antigen in solution to be used in selection rounds. This has proved useful in isolating high affinity binders from libraries (Hawkins *et al.*, 1992; Vaughan *et al.*, 1996).

It is possible to guide the selection process. For example, Tsui *et al.* (1996) were able to isolate an antibody against a second epitope of respiratory syncytial virus by adding an antibody against the dominant epitope into the panning solution, thereby blocking the dominant epitope. A novel method has been devised to select antibodies to cell surface proteins without the need to purify the antigen. This method utilizes antibodies or ligands that bind to the antigen. These are conjugated to horseradish peroxidase (HRP) and incubated together with a phage preparation from the library and the target antigen. On addition of biotin tyramine, and in the presence of hydrogen peroxide, phage in close proximity to the HRP-labelled antibody or ligand are covalently linked to biotin tyramine molecules. The labelled phage may then be collected (Osbourn *et al.*, 1998).

3.2.3 *Improving binding affinities of scFv after selection from a library*

The affinities of antibody fragments that are obtained from libraries may not be sufficient for their intended use. However, several methods of genetically modifiying V_H and V_L genes have been published. This has been done for antibody genes obtained from monoclonal antibodies (Thompson *et al.*, 1996) and from phage display libraries (Clackson *et al.*, 1991; Kang *et al.*, 1991a; Collet *et al.*, 1992; Gram *et al.*, 1992; Hawkins *et al.*, 1992; Marks *et al.*, 1992b; Barbas III *et al.*, 1994; Figini *et al.*, 1994; Yang *et al.*, 1995; Irving *et al.*, 1996; Low *et al.*, 1996; Schier *et al.*, 1996a, 1996b).

Random changes have been made by combining the V_H sequence of the original scFv with a repertoire of V_L sequences (or vice versa). Selection from the new repertoire has produced scFv with higher affinities than the original scFv (Clackson *et al.*, 1991; Kang *et al.*, 1991a; Collet *et al.*, 1992; Marks *et al.*, 1992b; Figini *et al.*, 1994). Other random changes are possible using error prone Taq polymerase in PCR (Gram *et al.*, 1992;

Hawkins *et al.*, 1992) and mutator strains of *E. coli* (Irving *et al.*, 1996; Low *et al.*, 1996). With these approaches the largest increases in affinity were observed for antibodies against hapten molecules (>100-fold; Low *et al.*, 1996; Marks *et al.*, 1992b).

Targeting mutations to the CDRs has been shown to be an effective technique for increasing antibody affinity. Random CDR sequences are produced by PCR using a mixture of degenerate oligonucleotide primers, and then incorporating these into the existing antibody sequence. Further selection against antigen is then done using the new antibody repertoire. So-called 'CDR walking' involves constructing a series of mutagenized repertoires. Each new repertoire is based on the previously selected high affinity binder. Yang *et al.* (1995) found results from this method to be consistent and produced Fab fragments against the envelope glycoprotein gp120 of human immunodeficiency virus type I with a 96-fold improvement in affinity over the original Fab fragment sequence. Also, individual changes can be made to the original antibody sequence and then each of the beneficial changes is then incorporated into a single sequence. This technique has proved to be less predictable, and not all combinations are additive. Yang *et al.* (1995) made changes in each of the CDR sequences separately and then tested pairwise combinations of these. Only one of the six combinations tested (improved V_L CDR1 with improved V_H CDR3) resulted in a Fab fragment with an affinity similar to that predicted from the two original improved sequences.

Parsimonious mutagenesis (Schier *et al.*, 1996a) is a technique in which mutagenic oligonucleotides are designed to minimize coding sequence redundancy and limit the number of amino acids that alter the structural features (for example, main chain conformation) of the existing antibody. Use of this method resulted in more than 50 per cent of new sequences tested having increased affinities over that of the original antibody sequence. However, the increases in affinity were less than one order of magnitude.

The largest increases in affinity (up to 1230-fold) have been achieved by mutagenesis of only the CDR3 sequences (Barbas III *et al.*, 1994; Yang *et al.*, 1995; Schier *et al.*, 1996b; Thompson *et al.*, 1996). This is probably because most of the high energy contact residues that are involved in the majority of the binding are located in the centre of the antibody-combining sites in the V_H and V_L CDR3 regions (Schier *et al.*, 1996b).

All of the above methods are very labour-intensive, as they involve the construction of new libraries. Often the most effective, least labour-intensive methods of improving the functional affinity of an antibody are to create multimers, which increases avidity. This can be done by expressing the antibody sequence in an appropriate expression vector (Plückthun and Pack, 1997), and is discussed in Chapter 4.

3.2.4 Antibodies obtained from phage display libraries

Phage display technology has been used to obtain antibodies against a wide range of antigens. These have included small haptens, proteins and viruses (Williamson *et al.*, 1993; Griffiths *et al.*, 1994; Nissim *et al.*, 1994; Vaughan *et al.*, 1996; Sheets *et al.*, 1998) and molecules on the surfaces of whole cells (Watters *et al.*, 1997; Osbourn *et al.*, 1998). Also, antibodies against plant pathogens and plant enzymes have been obtained from human scFv phage display libraries (Ziegler *et al.*, 1995, 1997; Harper *et al.*, 1997; Boonham and Barker, 1998; Griep *et al.*, 1998; Susi *et al.*, 1998) or from 'immunized' mouse scFv libraries (De Jaeger *et al.*, 1997; Nähring *et al.*, 1997) (Table 3.1). These have included scFv against *Blackcurrant reversion associated virus* (BRAV; genus *Nepovirus*). Previous attempts to raise antibodies to BRAV in rabbits produced antisera with a poor virus titre

Table 3.1 Plant pathogens and plant proteins for which scFv have been selected from phage display libraries

Antigen	Reference
Cucumber mosaic virus	Ziegler *et al.* (1995)
Potato leafroll virus	Harper *et al.* (1997)
Beet necrotic yellow vein virus	Griep *et al.* (1999a) Uhde and Koenig (unpublished data)
Potato virus Y	Boonham and Barker (1998) Uhde and Koenig (unpublished data)
Tomato spotted wilt virus	Griep and Schots (manuscript in preparation)
Blackcurrant reversion associated virus	Susi *et al.* (1998)
30 K movement protein of *Tobacco mosaic virus*	Nähring *et al.* (1997)
Dihydroflavonol 4-reductase (from *Petunia hybrida*)	De Jaeger *et al.* (1997)
Enoyl acyl carrier protein reductase (from *Brassica napus*)	Ziegler *et al.* (1997)
Lipopolysaccharide of *Ralstonia solanacearum* race 3	Griep *et al.* (1998)

and which contained antibodies to plant components, making them unsuitable for virus detection by ELISA (Susi *et al.*, 1998).

3.3 Anti-PLRV scFv from phage display libraries

3.3.1 Selection of antigen-binding phage

Two human scFv libraries (Nissim *et al.*, 1994; Vaughan *et al.*, 1996) were used as sources of scFv. Detailed protocols for selection are provided in Section 3.5.2. Selections were done against preparations of PLRV particles immobilized onto polystyrene tubes. Prior to immobilization, the particles were diluted in either phosphate-buffered saline (PBS), pH 7.4 or carbonate buffer, pH 9.6. PLRV particles remain intact in PBS but are disrupted in the carbonate buffer (Massalski and Harrison, 1987). Therefore, scFv selected against PLRV diluted in PBS were expected to recognize epitopes exposed on the virion surface (neotopes or metatopes), whereas those selected against PLRV diluted in carbonate buffer were expected to include those that bound to cryptotopes (see Chapter 1).

3.3.2 Assaying selected phage clones

After three to five rounds of selection against PLRV, individual clones were picked and tested for binding to PLRV in ELISA (see Section 3.5.2). The scFv were expressed in different ways for the tests:

- as phage antibody (by infecting *E. coli* TG-1 cells, on superinfection with M13KO7 helper phage);
- as soluble scFv (scFv-c-*myc*, in *E. coli* TG-1 cells, in the absence of glucose and in the presence of isopropyl β-D-thiogalactopyranoside); and
- as soluble scFv-*E. coli* alkaline phosphatase fusion proteins (scFv-AP) (after scFv genes from phage preparations were sub-cloned into the vector pSKAP/S (Griep *et al.*, 1999b) and expressed in *E. coli* TG-1 cells, on the addition of anhydrotetracycline).

Selections were done against PLRV particles from the smaller library (>10^8 members; Nissim *et al.*, 1994) and individual clones were tested for binding to PLRV as phage antibodies in ELISA. After three or four rounds of selection, more than 90 per cent of the clones tested by this method were positive, but none were successfully expressed as scFv-c-*myc* proteins. Two scFv genes, identified as phage antibodies, were then sub-cloned into soluble expression vectors (pDAP/2; pC_L) and scFv fusion proteins expressed for the detection of PLRV in ELISA (Harper *et al.*, 1997; Harper and Toth, unpublished data).

Selections against PLRV or disrupted PLRV virions were also made from the larger library (>10^{10} members; Vaughan *et al.*, 1996) and individual clones were tested for binding to PLRV as phage antibodies. Many positive binders were identified (>80 per cent of the clones tested), but could not be expressed as scFv-c-*myc* proteins. However, when one of these scFv genes was sub-cloned into pSKAP/S and pC_L, it was expressed as scFv-AP or scFv-C_L fusion proteins.

Other clones were identified by expressing individual scFv as scFv-c-*myc* proteins. Twenty per cent of scFv tested by this method bound PLRV in ELISA (Figure 3.2A). Six of these scFv genes were sub-cloned into pC_L for expressing scFv-C_L fusion proteins, but these could not be expressed as scFv-AP fusion proteins, when sub-cloned into pSKAP/S.

Furthermore, sub-cloning third and fourth round phage populations into pSKAP/S resulted in useful clones being identified. After three rounds of selection against PLRV, 30 per cent of the clones tested were positive (Figure 3.2B), and after four rounds of selection against disrupted PLRV, 60 per cent of the clones tested were positive. However, *Bst* NI fingerprinting and DNA sequence analyses indicated only two unique scFv sequences, one from each population of scFv-AP clones (Harper and Toth, unpublished data). Both of these scFv were sub-cloned into pC_L and expressed as scFv-C_L fusion proteins.

Thus, all three methods of expressing scFv for identifying antigen binders proved useful. PLRV-binding scFv were identified when scFv were expressed as phage antibodies, or, as scFv-c-*myc* proteins, but some of these sequences could not be expressed as functional scFv-AP fusion proteins. The third method, in which populations of scFv were sub-cloned into pSKAP/S and expressed individually as scFv-AP fusion proteins, was the most efficient method of obtaining useful scFv-AP fusion proteins.

3.3.3 *Detection of PLRV with anti-PLRV scFv*

The scFv genes from clones picked as phage antibodies or soluble scFv were sub-cloned into the pC_L expression vector for production of scFv-human kappa constant region (scFv-C_L) fusion proteins (McGregor *et al.*, 1994). Absorbance values (A_{405nm}) obtained

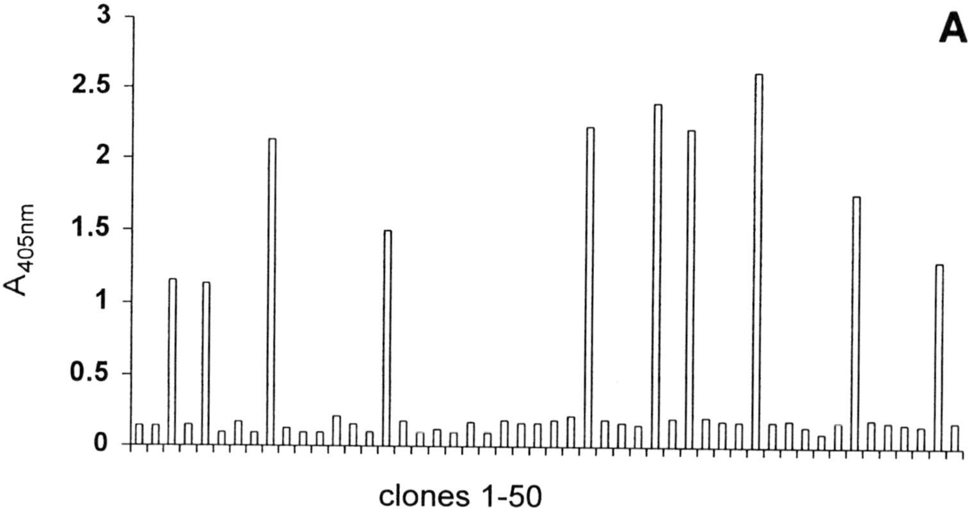

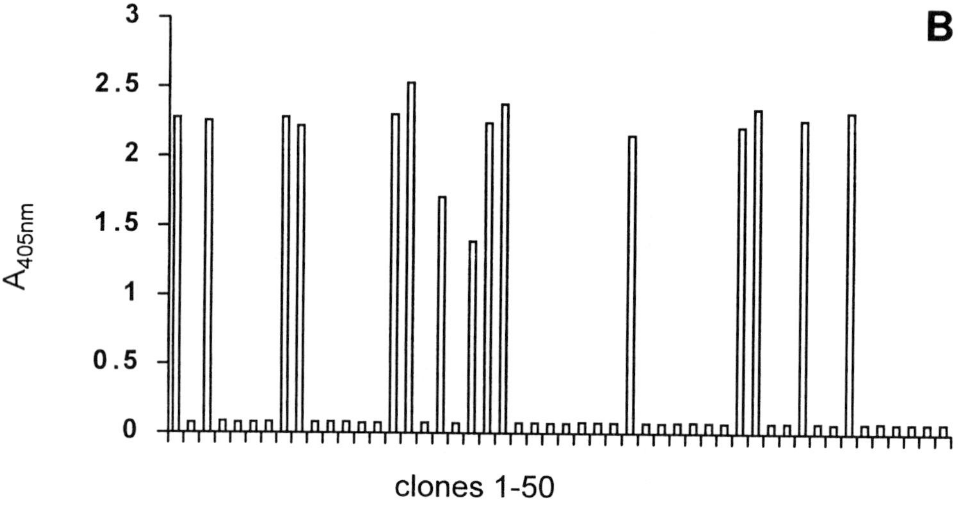

Figure 3.2 Detection of PLRV in ELISA by scFv obtained after three rounds of selection against PLRV. ScFv were expressed as soluble scFv-c-*myc* proteins (A) and ScFv-AP fusion proteins (B). ELISA plates were coated with 0.5 μg/ml of PLRV preparation diluted in PBS. Detection of binding was as described in Section 3.5.2. A_{405nm} values were obtained after overnight incubation with substrate buffer.

in ELISA tests with these scFv were greater for PLRV-infected samples when the scFv were expressed as scFv-C_L fusion proteins (A_{405nm} > 2.0) rather than as scFv-c-*myc* proteins (A_{405nm} = 0.1–0.2), illustrating the importance of the expression vectors used to produce useful detection reagents. ScFv-C_L and scFv-AP fusion proteins were used to detect PLRV in extracts of infected potato leaves by ELISA (Figure 3.3). A_{405nm} values were compared with those obtained in similar tests using the anti-PLRV monoclonal antibody, SCR3 (Massalski and Harrison, 1987). For some fusion proteins, the values obtained were greater than those obtained when using SCR3 (Figure 3.3A). Infected samples could be distinguished from non-infected samples within one hour of

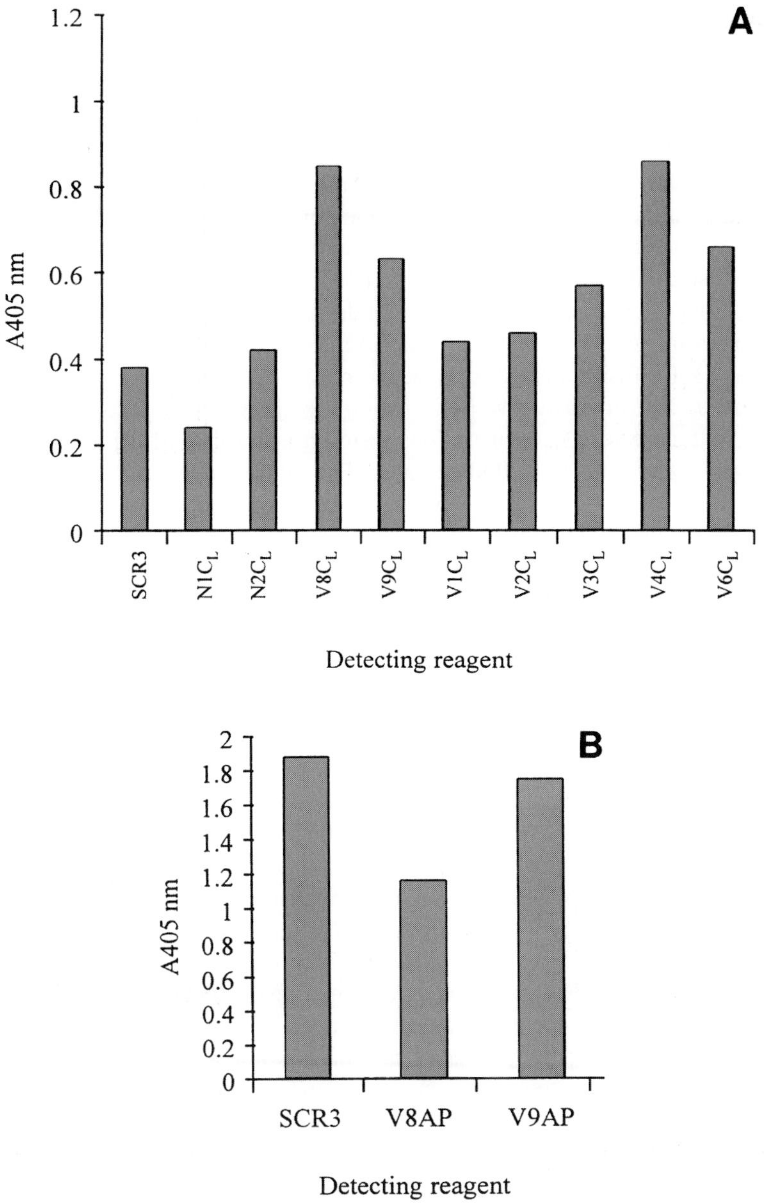

Figure 3.3 Detection of PLRV in potato leaf extracts with scFv-C_L (A) and scFv-AP (B) fusion proteins. A_{405nm} values were obtained after 40 minutes (A) or two hours (B) incubation with substrate buffer, and after subtraction of values obtained from non-infected control samples. N: scFv clones selected from a synthetic human scFv library (Nissim *et al.*, 1994); V: scFv clones selected from a 'non-immunized' human scFv library (Vaughan *et al.*, 1996).

adding substrate. Also, similar detection limits in ELISA (approximately five nanograms) were obtained (Harper and Toth, unpublished data). Thus, tests with the recombinant antibody fusion proteins gave comparable results to those obtained using a monoclonal antibody.

3.3.4 *Expression in* E. coli *and storage of scFv*

Chapter 4 gives detailed protocols for expressing scFv fusion proteins in *E. coli*, and for their subsequent purification. It is often possible to increase the yield of scFv-fusion proteins from shaker flask cultures by increasing the length of time for which expression is induced. For one scFv-AP fusion protein, most functional protein was detected in the culture medium and this continued to accumulate over a period of one week, at 16°C. The same scFv expressed as an scFv-C_L fusion protein was mainly detected in the periplasm and was harvested 40 hours post-induction, at 16°C (Harper and Toth, unpublished data). Therefore, it may be necessary to modify standard protocols for each scFv sequence and for each expression vector system, in order to maximize yields. This way, several milligrams of each scFv-fusion protein may be expressed per litre of culture fluid. Typically, by our methods, 2 ml of purified scFv-C_L fusion protein were obtained from every 500 ml of bacterial culture fluid. The preparations were diluted 100-fold for use in ELISA, providing enough material to test 2000 samples. The undiluted scFv-C_L were stable at 4°C for at least six months. In contrast, preparations of scFv-AP fusion proteins were found to be unstable after purification and storage at 4°C in PBS, and therefore crude cell extracts containing scFv-AP fusion protein were used as detection reagents. These were stable for approximately two weeks at 4°C and were diluted tenfold for ELISA. All constructs tested so far were active when reconstituted after lyophilization.

3.4 Conclusions

Phage display technology can be used to produce high affinity reagents with binding properties similar to those of monoclonal antibodies produced from immunized animals and is particularly useful for obtaining reagents against weak immunogens. A panel of scFv against PLRV was isolated from large 'non-immunized' human scFv libraries. The scFv were genetically modified to produce scFv-fusion proteins, suitable for use in assays to detect PLRV in infected potato leaf extracts. Since 1995, scFv against at least ten plant pathogens and plant proteins have been obtained from phage display scFv antibody libraries. There is potential for wide scale application of phage display technology in the plant sciences industry and the provision of antibody genes for plant expression is discussed in other chapters of this book. However, more widespread application of phage display technology may be limited by the availability of, or ready access to, large, high quality libraries. Few such libraries exist at present. They are commercially valuable and priority is given to medical research. Construction of such libraries is extremely labour-intensive and strategies for optimizing the production of very large libraries are being sought continually (Sheets *et al.*, 1998). In the future, it may be necessary to construct a phage display library for general use in plant science research, with funding and resources contributed by, for example, a European Union consortium.

3.5 Reagents and protocols

3.5.1 *Reagents*

The following reagents are required for the protocols described in Section 3.5.2:

LB medium (per litre): 10 g tryptone, 5 g yeast extract, 5 g NaCl

LB AG: LB medium supplemented with 100 μg/ml ampicillin and 1 per cent glucose (w/v)

2 × TY medium (per litre): 16 g tryptone, 10 g yeast extract, 5 g NaCl

SOB medium (per litre): 20 g tryptone, 5 g yeast extract, 0.5 g NaCl, 2.5 g KCl, supplemented with 10 ml of 1 M $MgCl_2$, 10 ml of 1 M $MgSO_4$ and 17 ml of 40 per cent glucose (w/v)

LB agar: LB medium containing 1.5 per cent bacto agar

TYE AG agar (per litre): 15 g bacto agar, 8 g NaCl, 10 g tryptone, 5 g yeast extract, 100 μg/ml ampicillin (w/v), 1 per cent glucose (w/v)

50 mM carbonate buffer, pH 9.6: (per litre: 1.59 g Na_2CO_3, 2.93 g $NaHCO_3$)

PBS: Phosphate-buffered saline, pH 7.4 (per litre: 8 g NaCl, 0.2 g KH_2PO_4, 2.9 g $Na_2HPO_4.12.H_2O$, 0.2 g KCl)

MPBS: PBS containing 3 per cent low fat milk powder (w/v) (Marvel, dried skimmed milk)

PBS Tween: PBS containing 0.1 per cent (v/v) polyoxyethylenesorbitan monolaurate (Tween 20)

Extraction buffer for tests on Physalis floridana *or potato leaves*: PBS containing 1 per cent polyvinylpyrrolidone (w/v) and 0.05 per cent Tween 20 (v/v)

100 mM Triethylamine, pH 12 (TEA)

1 M TrisHCl, pH 7.4

PEG/NaCl solution: 2.5 M NaCl aqueous solution containing 20 per cent (w/v) polyethylene glycol 6000

E. coli *TG-1 cells*: maintained on minimal media plates supplemented with thiamine

3.5.2 **Protocols**

Selecting anti-PLRV scFv from a scFv library

- To prepare antigen-coated tube, dilute PLRV preparation (5–50 μg) in 1 ml of PBS or carbonate buffer, add to immunotube (Nunc MaxiSorp) and incubate overnight at room temperature.
- Rinse tube three times with PBS.
- Fill tube with MPBS and incubate for two hours at 37°C.
- In an Eppendorf microcentrifuge tube, pre-incubate phage by adding 200 μl of 6 × MPBS to 1 ml of phage preparation suspended in PBS (approximately 1×10^{12} from the previous selection round, or from the original library) and incubate at room temperature for one hour.
- Remove blocking buffer from the antigen-coated tube and rinse three times with PBS.
- Add pre-incubated phage to the blocked antigen-coated tube and incubate for one hour at 37°C.
- Remove phage solution and rinse the tube 20 times with PBS Tween and then 20 times with PBS.

- Add 1 ml of TEA and incubate at room temperature for 10 minutes.
- Neutralize the eluted phage by transferring into a clean Eppendorf microcentrifuge tube containing 500 µl of 1 M Tris/HCl, pH 7.4 and store on ice until required for infection of *E. coli* cells.
- Add 200 µl of 1 M Tris/HCl, pH 7.4 to the empty antigen-coated tube and store on ice until required for infection of *E. coli* cells.

Infection of E. coli *TG-1 cells with eluted phage*

- Add half of the neutralized phage preparation to 15 ml of log phase TG-1 *E. coli* cells, also, add 3 ml of log phase *E. coli* TG-1 cells to the antigen-coated tube and incubate for 30 minutes at 37°C.
- Combine the two sources of infected TG-1 cells and incubate for a further 30 minutes at 37°C with shaking.
- Spread 10 µl and 100 µl aliquots onto TYE AG plates (titre plates) and incubate overnight at 30°C.
- Sediment the remainder of the cells by centrifugation (3500 **g** for ten minutes), resuspend in 500 µl of 2 × TY media, spread onto TYE AG plates (output plates) and incubate overnight at 30°C.
- Count colonies on the titre plates and calculate phage output.

Amplification of output phage

- Scrape off colonies from output plates into 100 ml of 2 × TY AG medium and incubate with shaking for two hours at 37°C.
- Transfer 10 ml into a sterile 50 ml polypropylene centrifuge tube, add M13 K07 helper phage (at least tenfold excess of helper phage to *E. coli* cells) and incubate for 30 minutes (stationary) followed by a further 30 minutes with shaking, at 37°C.
- Sediment the cells (3500 **g** for ten minutes), resuspend in 100 µl of 2 × TY medium containing 100 µg/ml ampicillin and 50 µg/ml kanamycin and shake overnight at 30°C.

Harvesting phage preparation

- Sediment the *E. coli* cells in two sterile 50 ml polypropylene centrifuge tubes (3500 **g** for 30 minutes), transfer the supernatant to sterile tubes and add one fifth volume of PEG/NaCl solution to precipitate the phage particles and incubate on ice for at least 45 minutes.
- Sediment the phage particles (3500 **g** for 45 minutes).
- Resuspend pellet in 1 ml of PBS, transfer to an Eppendorf microcentrifuge tube and precipitate the phage particles for a second time; add 200 µl of PEG/NaCl solution, incubate on ice for 20 minutes, sediment phage (17 000 **g** for 20 minutes) and resuspend in 1 ml PBS.
- Sediment any cellular debris (17 000 **g** for 30 seconds), transfer phage particles in supernatant to a fresh Eppendorf tube and store at 4°C (with 0.05 per cent sodium azide (w/v)).
- Titre phage and use for further selection against antigen.

Phage titres (colony forming units)

- Label Eppendorf microcentrifuge tubes 1, 2, 3 and 4 and add 1 ml of LB medium to tubes 1, 2 and 4.
- Add 10 μl of phage to tube 1, mix and add 10 μl of the diluted phage to tube 2 and mix.
- In tube 3, mix 1 ml of log phase *E. coli* TG-1 cells and 10 μl of diluted phage from tube 2.
- Incubate for 30 minutes at 37°C.
- Add 10 μl of the phage/cell mix to tube 4 and mix.
- Plate 10 μl and 100 μl aliquots from tube 4 onto TYE AG plates and incubate overnight at 30°C.
- Each colony on the '10 μl' plate corresponds to 1×10^{10} colony forming units (cfu).

Preparation of M13 K07 helper phage

(original stocks may be purchased from Pharmacia Biotech)

- Inoculate 5 ml of log phase *E. coli* TG-1 cells with 3–5 plaques from previous helper phage titre plate (may be stored on plates for several months).
- Incubate at 37°C for 30 minutes and a further 30 minutes with shaking.
- Transfer culture to shaker flask containing 100 ml of LB supplemented with 50 μg/ml of kanamycin.
- Shake overnight at 37°C.
- Precipitate with PEG/NaCl as for output phage.
- Titre helper phage (as for output phage except the phage are plated in top agar, as follows: 2 ml SOB medium containing 14 mg agar at 42°C are mixed with 100 μl of an overnight culture of *E. coli* TG-1 cells, and 10 μl of phage from tube 4 is added; the mixture is then poured onto an LB plate, allowed to set and incubated overnight at 37°C).

Bst *N1 fingerprinting (to check the diversity of clones)*

- Make a PCR mix on ice, for the appropriate number of samples (for example, 200 μl for 10 × 20 μl volume samples:

10 × Taq polymerase buffer	20 μl
5 mM dNTP mixture	8 μl
10 pmol/μl primer 1*	10 μl
10 pmol/μl primer 2**	10 μl
Distilled water	150 μl
Taq DNA polymerase (10 Units)	2 μl

*Primer 1: 5′ A G C G G A T A A C A A T T T C A C A C A G G 3′;
**Primer 2: 5′ G T C G T C T T T C C A G A C G T T A G T 3′

- Aliquot 20 μl volumes of the mix into 0.5 ml PCR tubes and add 0.5 μl of plasmid DNA samples.

- Amplify DNA by PCR (30 cycles of: 94°C, 1 minute; 55°C, 1 minute; 72°C, 2 minutes; followed by 72°C, 10 minutes).
- Ethanol precipitate the PCR products.
- Resuspend the DNA and digest with *Bst* NI restriction enzyme (NEB) (20 μl volumes, each containing 100 μg/ml BSA and 2–5 units of *Bst* NI) for two hours at 60°C.
- Resolve the digestion products on a 3 per cent agarose gel and visualize by staining with ethidium bromide.

Oligonucleotide primers for sequencing scFv clones

- For sequencing V_H genes derived from the pCANTAB series (Vaughan and Nissim libraries) the primer LMB 3 is used: 5′ C A G G A A A C A G C T A T G A C C A 3′.
- For sequencing the V_L genes from the pCANTAB series (Vaughan library) the primer myc seq 10 is used: 5′ C T C T T C T G A G A T G A G T T T T T G 3′.

Expression of phage antibodies for ELISA

- Inoculate 5 ml of LB AG with a single colony and incubate at 30°C overnight with shaking.
- Inoculate 5 ml of 2 × TY AG medium with 500 μl of the overnight culture, add helper phage to a final concentration of 5×10^{10} pfu/ml (a tenfold excess of helper phage to *E. coli* cells) and incubate for two hours at 37°C with shaking.
- Sediment cells from a 1 ml aliquot of the *E. coli*/helper phage culture and resuspend in 5 ml of 2 × TY containing 100 μg ampicillin and 50 μg/ml kanamycin.
- Incubate at 30°C overnight with shaking.

Harvesting phage antibodies for ELISA

- Aliquot 1 ml of each overnight culture into Eppendorf microcentrifuge tubes and sediment cells (17 000 **g** for two minutes).
- Transfer culture to clean tubes, add 200 μl of PEG/NaCl and incubate on ice for 30 minutes before sedimenting the phage (17 000 **g** for 20 minutes).
- Resuspend pellet in 200 μl MPBS.

Expression of scFv-alkaline phosphatase fusion proteins for ELISA

- Inoculate 5 ml of LB AG medium with a single colony and incubate at 30°C overnight with shaking.
- Inoculate 5 ml of 2 × TY AG medium with 50 μl of the overnight culture and incubate for three hours at 30°C, with shaking, followed by one hour at 16°C, with shaking.
- Add fresh ampicillin, anhydrotetracycline and glycerol to final concentrations of 100 μg/ml, 0.2 μg/ml and 1 per cent, respectively.
- Incubate overnight at 16°C, with shaking.

*Expression of scFv-c-*myc *fusion proteins in 96-well polystyrene plates for ELISA*

- Inoculate 100 μl of LB AG medium with a single colony and incubate at 30°C overnight with shaking.
- Inoculate 100 μl of 2 × TY AG medium with 1 μl of the overnight culture and incubate for four hours at 30°C, with shaking.
- Sediment the cells and resuspend in 100 μl of 2 × TY medium containing 100 μg/ml ampicillin (w/v) and 1 mM IPTG.
- Incubate overnight at 30°C with shaking.

*Harvesting soluble scFv from the periplasm (scFv-c-*myc *and scFv-AP fusion proteins)*

- Sediment cells (3500 **g** for ten minutes) and resupend in one-twentieth volume of 30 per cent sucrose (w/v) containing 50 mM Tris/HCl, pH 8.0 and 1 mM ethylenediaminetetra-acetic acid disodium salt (EDTA), pH 8.0.
- Sediment cells as above and resuspend in the same volume of 5 mM $MgSO_4$ and incubate on ice for 45 minutes.
- Sediment cell debris as above and collect soluble periplasmic extract.

ELISA

- Coat each ELISA plate well with 100 μl of anti-PLRV polyclonal antibody (1 μg/ml in carbonate buffer) and incubate for four hours at room temperature or overnight at 4°C.
- Rinse wells three times with PBS, add 200 μl of MPBS to each well and incubate at 30°C for one hour.
- Rinse wells three times with PBS and add plant extracts (one in five dilutions (w/v) of virus-infected, or non-infected negative control samples, in plant extract buffer) and incubate overnight at 4°C.
- Rinse wells three times with PBS Tween and three times with PBS.
- Add scFv preparations and incubate for one hour at 30°C.
- Rinse wells as above and then:
  - For phage ELISA: add MPBS containing 1 μg/ml anti-M13 antibody–alkaline phosphatase conjugate, incubate for one hour at 30°C, rinse wells as above and add alkaline phosphatase substrate (1 mg/ml *p*-nitrophenyl phosphate, disodium (w/v) in 10 per cent diethanolamine (v/v), pH 9.8).
  - For detecting scFv-c-*myc* fusion proteins: add MPBS containing 1 μg/ml anti-c-*myc* monoclonal antibody 9E10 (Invitrogen) and 1 μg/ml anti-mouse-IgG–alkaline phosphatase conjugate (Sigma), incubate for one hour at 30°C, rinse wells as above and add alkaline phosphatase substrate.
  - For detection of scFv-alkaline phosphatase fusion proteins: add alkaline phosphatase substrate.
- For each scFv sample, compare absorbance values for infected and non-infected plant extracts.

Acknowledgements

We thank the MRC Centre for Protein Engineering, and Cambridge Antibody Technology for access to the human scFv libraries. This work was funded by EU contract AIR3 CT94-1046 and the Scottish Office Agriculture, Environment and Fisheries Department.

References

BARBAS III, C.F., KANG, A.S., LERNER, R.A. and BENKOVIC, S.J., 1991, Assembly of combinatorial libraries on phage surfaces. The gene III site, *Proceedings of the National Academy of Science USA*, **88**, 7978–7982.

BARBAS III, C.F., HU, D., DUNLOP, N., SAWYER, L., CABABA, D., HENDRY, R.M., *et al.*, 1994, *In vitro* evolution of a neutralising human antibody to human immunodeficiency virus type 1 to enhance affinity and broaden strain cross-reactivity, *Proceedings of the National Academy of Science USA*, **91**, 3809–3813.

BOONHAM, N. and BARKER, I., 1998, Strain specific recombinant antibodies to Potato virus Y potyvirus, *Journal of Virological Methods*, **74**, 193–199.

CLACKSON, T., HOOGENBOOM, H.R., GRIFFITHS, A.D. and WINTER, G., 1991, Making antibody fragments using phage display libraries, *Nature*, **352**, 624–628.

CLARK, M.F., 1992, Immunodiagnostic techniques for plant mycoplasma-like organisms, in *Techniques for the Rapid Detection of Plant Pathogens*, eds Duncan, J.M. and Torrance, L., Blackwell Scientific Publications, Oxford, pp. 34–45.

COLLET, T.A., ROBEN, P., O'KENNEDY, R., BARBAS III, C.F., BURTON, D.R. and LERNER, R.L., 1992, A binary plasmid system for shuffling combinatorial antibody libraries, *Proceedings of the National Academy of Science USA*, **89**, 10026–10030.

DE JAEGER, G., BUYS, G., EECKHOUT, D., BRUYNS, A.M., DENEVE, M., DEWILDE, C., *et al.*, 1997, Use of phage display for isolation and characterisation of single chain variable fragments against dihydroflavonol 4-reductase from Petunia hybrida, *FEBS Letters*, **403**, 116–122.

DEWEY, F.M., 1992, Detection of plant-invading fungi by monoclonal antibodies, in *Techniques for the Rapid Detection of Plant Pathogens*, eds Duncan, J.M. and Torrance, L., Blackwell Scientific Publications, Oxford, pp. 47–62.

FIGINI, M., MARKS, J.D., WINTER, G. and GRIFFITHS, A.D., 1994, *In vitro* assembly of repertoires of antibody chains on the surface of phage by renaturation, *Journal of Molecular Biology*, **239**, 68–78.

GHERARDI, E. and MILSTEIN, C., 1992, Original and artificial antibodies, *Nature*, **357**, 201–202.

GHERARDI, E. and WINTER, G., 1994, Isolation of high affinity antibodies directly from large synthetic repertoires, *EMBO Journal*, **13**, 3245–3260.

GRAM, H., MARCONI, L.-A., BARBAS III, C.F., COLLET, T.A., LERNER, R.A. and KANG, A.S., 1992, *In vitro* selection and affinity maturation of antibodies from a naive combinatorial immunoglobulin library, *Proceedings of the National Academy of Science USA*, **89**, 3576–3580.

GRIEP, R.A., VAN TWISK, C., VAN BECKHOVEN, J.R.C.M., VAN DER WOLF, J.M. and SCHOTS, A., 1998, Development of specific recombinant monoclonal antibodies against the lipopolysaccharide of *Ralstonia solanacearum* race 3, *Phytopathology*, **88**, 795–803.

GRIEP, R.A., VAN TWISK, C. and SCHOTS, A., 1999a, Selection of beet necrotic yellow vein virus specific monoclonal antibodies from a semi-synthetic combinatorial antibody library, *European Journal of Plant Pathology* (in press).

GRIEP, R.A., VAN TWISK, C., KERSCHBAUMER, R.J., HARPER, K., TORRANCE, L., HIMMLER, G., *et al.*, 1999b, pSKAP/S: an expression vector for the production of single-chain Fv alkaline phosphatase fusion-proteins, *Protein, Purification and Expression* (in press).

GRIFFITHS, A.D., MALMQVIST, M., MARKS, J.D., BYE, J.M., EMBLETON, M.J., MCCAFFERY, J., *et al.*, 1993, Human anti-self antibodies with high specificity from phage display libraries, *EMBO Journal*, **12**, 725–734.

GRIFFITHS, A.D., WILLIAMS, S.C., HARTLEY, O., TOMLINSON, I.M., WATERHOUSE, P., CROSBY, W.L., *et al.*, 1994, *EMBO Journal*, **13**, 3245–3260.

HAMERS-CASTERMAN, C., ATARHOUCH, T., MUYLDERMANS, S., ROBINSON, G., HAMERS, C., SONGA, E.B., *et al.*, 1993, Naturally-occurring antibodies devoid of light chains, *Nature*, **363**, 446–448.

HARPER, K., KERSCHBAUMER, R.J., ZIEGLER, A., MACINTOSH, S.M., COWAN, G.H., HIMMLER, G., *et al.*, 1997, A scFv-alkaline phosphatase fusion protein which detects potato leafroll luteovirus in plant extracts by ELISA, *Journal of Virological Methods*, **63**, 237–242.

HARRISON, J.L., WILLIAMS, S.C., WINTER, G. and NISSIM, A., 1996, Screening of phage antibody libraries, *Methods in Enzymology*, **267**, 83–109.

HAWKINS, R.E., RUSSELL, S.J. and WINTER, G., 1992, Selection of phage antibodies by affinity binding mimicking affinity maturation, *Journal of Molecular Biology*, **226**, 889–896.

HAWKINS, R.E., RUSSELL, S.J., BAIER, M. and WINTER, G., 1993, The contribution of contact and non-contact residues of antibody in the affinity of binding to antigen. The interaction of mutant D1.3 antibodies with lysosyme, *Journal of Molecular Biology*, **234**, 958–964.

HOOGENBOOM, H.R., GRIFFITHS, A.D., JOHNSON, K.S., CHISWELL, D.J., HUDSON, P. and WINTER, G., 1991, Multi-subunit proteins on the surface of filamentous phage: methodologies for displaying antibody (Fab) heavy and light chains, *Nucleic Acid Research*, **19**, 4133–4137.

IRVING, R.A., KORTT, A.A. and HUDSON, P.J., 1996, Affinity maturation of recombinant antibodies using *E. coli* mutator cells, *Immunotechnology*, **2**, 127–143.

JOHNSON, K.S. and HAWKINS, R.E., 1996, Affinity maturation of antibodies using phage display, in *Antibody Engineering: A Practical Approach*, eds J. McCafferty, H.R. Hoogenboon and D.J. Chiswell (A Practical Approach series; 169), Oxford University Press, Oxford, pp. 41–58.

JOHNSON, G., KABAT, E.A. and WU, T.T., 1996, Kabat database of sequences of proteins of immunological interest, in *Weir's Handbook of Experimental Immunology*, fifth edition, eds L.A. Hersberg, W.M. Weir and C. Blackwell, Blackwell Science Inc., Cambridge, MA, pp. 6.1–6.21.

KANG , A.S., JONES, T.M. and BURTON, D.R., 1991a, Antibody redesign by chain shuffling from random combinatorial libraries, *Proceedings of the National Academy of Science USA*, **88**, 11120–11123.

KANG, A.S., BARBAS III, C.F., JOUDA, K.D., BENKOVIC, S.J. and LERNER, R.A., 1991b, Linkage of recognition and replication functions by assembling combinatorial antibody Fab libraries along phage surfaces, *Proceedings of the National Academy of Science USA*, **88**, 4363–4366.

KRAMER, K., 1998, Synthesis of pesticide-specific single-chain F_V utilizing the recombinant phage antibody system (RPAS, Pharmacia). A detailed protocol, *Analytical Letters*, **31**, 67–92.

KREBS, B., PACK, P., GE, L., VIRNEKAS, B., WELLNHOFER, G., HOESS, A., *et al.*, 1998, HuCAL: modular human combinatorial antibody libraries. Recombinant antibodies for pioneering therapeutics, International Conference, 17–19 June 1998, Arlington, VA (Meeting Abstract).

LOW, N.M., HOLLIGER, P. and WINTER, G., 1996, Mimicking somatic hypermutation: affinity maturation of antibodies displayed on bacteriophage using a bacterial mutator strain, *Journal of Molecular Biology*, **260**, 359–368.

MARKS, J.D., HOOGENBOOM, H.R., BONNERT, T.P., MCCAFFERTY, J., GRIFFITHS, A.D. and WINTER, G., 1991, By-passing immunization: human antibodies from V-gene libraries displayed on phage, *Journal of Molecular Biology*, **222**, 581–597.

MARKS, J.D., HOOGENBOOM, H.R., GRIFFITHS, A.D. and WINTER, G., 1992a, Molecular evolution of proteins on filamentous phage, *Journal of Biological Chemistry*, **267**, 16007–16010.

MARKS, J.D., GRIFFITHS, A.D., MALMQVIST, M., CLACKSON, T., BYE, J.M and WINTER, G., 1992b, By-passing immunization: building high affinity human antibodies by chain shuffling, *Bio/Technology*, **10**, 779–783.

MASSALSKI, P.R. and HARRISON, B.D., 1987, Properties of monoclonal antibodies to potato leafroll luteovirus and their use to distinguish virus isolates differing in aphid transmissibility, *Journal of General Virology*, **68**, 1813–1821.

McCafferty, J., Griffiths, A.D., Winter, G. and Chiswell D.J., 1990, Phage display antibodies: filamentous phage displaying antibody variable domains, *Nature*, **348**, 552–554.

McGregor, D.P., Molloy, P.E., Cunningham, C. and Harris, W.J., 1994, Spontaneous assembly of bivalent single chain antibody fragments in *Escherichia coli*, *Molecular Immunology*, **31**, 219–226.

McGuinness, B.T., Walter, G., FitzGerald, K., Schuler, P., Mahoney, W., Duncan, A.R. and Googenboom, H.R., 1996, Phage diabody repertoires for selection of large numbers of bispecific antibody fragments, *Nature Biotechnology*, **14**, 1149–1154.

Nähring, J.M., Schummer, G., Lewandowski, D.J., Liao, Y.-C. and De Jaeger, G., 1997, Isolation and characterization of scFv-fragments binding to the 30K movement protein of TMW by phage display, Perspectives on Protein Engineering, 6th International Conference, 28 June–1 July 1997, Norwich, UK, Abstract 102.

Nissim, A., Hoogenboom, H.R., Tomlinson, I.M., Flynn, G., Midgley, C., Lane, D. and Winter, G., 1994, Antibody fragments from a single pot phage display library as immunological reagents, *EMBO Journal*, **13**, 692–698.

Osbourn, J.K., Derbyshire, E.J., Vaughan, T.J., Field, A.W. and Johnson, K.S., 1998, Pathfinder selection: *in situ* isolation of novel antibodies, *Immunotechnology*, **3**, 293–302.

Plückthun, A. and Pack, P., 1997, New protein engineering approaches to multivalent and bispecific antibody fragments, *Immunotechnology*, **3**, 83–105.

Pope, A., Pritchard, K., Williams, A., Roberts, A., Hackett, J.R., MacCafferty, J. and Johnson, K.S., 1996, *In vitro* selection of a high affinity antibody to oestradiol using a phage display human library, *Immunotechnology*, **2**, 209–217.

Ridder, R., Schmitz, R., Legay, F. and Gram, H., 1995, Generation of rabbit monoclonal antibody fragments from a combinatorial phage display library and their production in the yeast *Pichia pastoris*, *Bio/Technology*, **13**, 255–260.

Schier, R., Balint, R.F., McCall, A., Apell, G., Larrick, J.W. and Marks, J.D., 1996a, Identification of functional and structural amino-acid residues by parsimonious mutagenesis, *Gene*, **169**, 147–155.

Schier, R., McCall, A., Adams, G.P., Marchall, K.W., Merritt, H., Yim, M., *et al.*, 1996b, Isolation of picomolar affinity anti-c-erbB-2 single-chain Fv by molecular evolution of the complementarity determining regions in the center of the antibody binding site, *Journal of Molecular Biology*, **263**, 551–557.

Sheets, M.D., Amersdorfer, P., Finnern, R., Sargent, P., Lindqvist, E., Schier, R., *et al.*, 1998, Efficient construction of a large non-immune phage antibody library: the production of high-affinity single-chain antibodies to protein antigens, *Proceedings of the National Academy of Science USA*, **95**, 6157–6162.

Stead, D.E., 1992, Techniques for detecting and identifying plant pathogenic bacteria, in *Techniques for the Rapid Detection of Plant Pathogens*, eds Duncan, J.M. and Torrance, L., Blackwell Scientific Publications, Oxford, pp. 76–111.

Susi, P., Ziegler, A. and Torrance, L., 1998, Selection of single chain variable fragment antibodies to blackcurrant reversion associated virus from a synthetic phage display library, *Phytopathology*, **88**, 230–233.

Thompson, J., Pope, T., Tung, J.-S., Chan, C., Hollis, G., Mark, G. and Johnson, S., 1996, Affinity maturation of a high-affinity human monoclonal antibody against the third hypervariable loop of human immunodeficiency virus: use of phage display to improve affinity and broaden strain reactivity, *Journal of Molecular Biology*, **256**, 77–88.

Tomlinson, I.M., Williams, S.C., Corbett, S.J., Cox, J.P.L. and Winter, G., 1996, *V Base Sequence Directory*, MRC Centre for Protein Engineering, Cambridge, UK (www.mrc-cpe.cam.ac.uk/imt-doc/vbase-home-page.html).

Torrance, L., 1992, Serological methods to detect plant viruses: production and use of monoclonal antibodies, in *Techniques for the Rapid Detection of Plant Pathogens*, eds Duncan, J.M. and Torrance, L., Blackwell Scientific Publications, Oxford, pp. 7–33.

Torrance, L., 1995, Use of monoclonal antibodies in plant pathology, *European Journal of Plant Pathology*, **101**, 351–363.

TSUI, P., TORNETTA, M.A., AMES, R.S., BANKOSKY, B.C., GRIEGO, S., SILVERMAN, C., *et al.*, 1996, Isolation of a neutralising human RSV antibody from a dominant, non-neutralising immune repertoire by epitope-blocked panning, *Journal of Immunology*, **157**, 772–780.

VAUGHAN, T.J., WILLIAMS, A.J., PRITCHARD, K., OSBOURN, J.K., POPE, A.R., EARNSHAW, J.C., *et al.*, 1996, Human antibodies with sub-nanomolar affinities isolated from a large non-immunized phage display library, *Nature Biotechnology*, **14**, 309–314.

WATTERS, J.M., TELLEMAN, P. and JUNGHANS, R.P., 1997, An optimized method for cell-based phage display panning, *Immunotechnology*, **3**, 21–29.

WILLIAMSON, R.A., BURIONI, R., SANNA, P.P., PARTRIDGE, L.J., BARBAS III, C.F. and BURTON, D.R., 1993, Human monoclonal antibodies against a plethora of viral antigens from single combinatorial libraries, *Proceedings of the National Academy of Science USA*, **90**, 4141–4145.

WINTER, G. and MILSTEIN, C., 1991, Man-made antibodies, *Nature*, **349**, 293–299.

WINTER, G., GRIFFITHS, A.D., HAWKINS, R.E. and HOOGENBOOM, H.R., 1994, Making antibodies by phage display technology, *Annual Review of Immunology*, **12**, 433–455.

YANG, W.-P., GREEN, K., PINZ-SWEENEY, S., BRIONES, A.T., BURTON, D.R. and BARBAS III, C.F., 1995, CDR walking mutagenesis for the affinity maturation of a potent human anti-HIV-1 antibody into the picomolar range, *Journal of Molecular Biology*, **254**, 392–403.

ZIEGLER, A., TORRANCE, L., MACINTOSH, S.M., COWAN, G.H. and MAYO, M.A., 1995, Cucumber mosaic cucumovirus antibodies from a synthetic phage display library, *Virology*, **214**, 235–238.

ZIEGLER, A., MACINTOSH, S.M., TORRANCE, L., SIMON, W. and SLABAS, A.R., 1997, Recombinant antibody fragments that detect enoyl acyl carrier protein reductase in *Brassica napus*, *Lipids*, **32**, 805–809.

4

Dedicated Expression Vectors for the Production of Diagnostic Reagents

RANDOLF J. KERSCHBAUMER AND GOTTFRIED HIMMLER

4.1 Introduction

In recent years, a great variety of antigen-binding proteins has been produced in *Escherichia coli* (Skerra, 1993; Plückthun and Pack, 1997), several with the potential to be used in immunoassays such as Western blotting or enzyme-linked immunosorbent assay (ELISA). Recombinant immunological reagents for detecting plant pathogens may be developed using antibody fragment genes, derived by the standard cloning and selection methods described in detail in Chapters 2 and 3 of this book. Thus, recombinant antibody technology has great potential in plant pathology and agriculture for designing and producing novel diagnostic tools, which may replace conventional immunoassay reagents produced by immunization of animals or by hybridoma cultures.

Many routine immunological assays involved in assessing plant health are still based on polyclonal sera obtained from immunized animals. The technology is well established for the production and purification of such reagents and has been optimized for many different antigens (Van Regenmortel, 1982). To produce sera with the desired properties, over a long period of time, animals must be immunized repeatedly. Therefore, preparation of large quantities of antigen is necessary, which may be costly and time consuming. Furthermore, production of polyclonal sera is complicated by the requirement of animal house facilities. The efforts necessary may increase dramatically in the case of large-scale production of antisera, achieved by immunizing larger animals, such as sheep or goats.

Other sources of antibodies are hybridoma cell lines, producing antigen-specific monoclonal antibodies. Since first being described by Köhler and Milstein in 1975, monoclonal antibody technology has been applied in many laboratories throughout the world. Hybridoma technology involves immunizing animals (usually mice), fusing their lymphocytes with myeloma cells, followed by screening for stable hybridoma cell lines expressing antibodies with the desired properties. A stable, monoclonal cell line is an inexhaustible source of a defined reagent of constant quality. Immunization is only necessary at the beginning of the process and in comparison with polyclonal antibody production, little antigen is needed. Cell culture facilities such as laminar flow hoods, carbon dioxide incubators, and liquid nitrogen storage facilities are required for production, maintainance and storage of hybridomas. Large-scale production of monoclonal antibodies is possible by fermentation (reviewed by

Bibila and Robinson, 1995) but this is troublesome due to slow growing rates, low cell densities and cell sensitivity. Conditions for each cell line need to be optimized and high protein-levels (>1 g/l culture medium) are achieved only in exceptional cases. Hybridomas may also be cultured as ascitic fluid in mice, but this technique is particularly unpleasant for the animals and also suffers from drawbacks such as poor reproducibility.

Recombinant antibody technology may overcome many of the problems associated with producing antibodies by immunization of animals or by hybridoma cultures. Genes of antibody fragments, which may be expressed in *E. coli*, are developed either by subcloning antibody genes from hybridoma cells (Chapter 2 of this book) or by library selection utilizing phage display technology (Chapter 3 of this book). Insertion of an antibody fragment gene into an appropriate expression vector and transformation into *E. coli* provides a source of a well defined reagent, which can be sequenced and characterized. Bacteria harbouring the vector construct can be stored at –70°C, and the reagent may be re-expressed reproducibly at any time. Bacterial cells are in general easier and less time consuming to handle than animal cells and require, in comparison, little specialized equipment. Moreover, fermentation of bacteria for large-scale production tends to be quicker and much less problematic than for hybridoma cell lines.

In this chapter we describe different types of antigen-binding proteins that may be produced in bacteria, and how these may be used as immunological diagnostic reagents. Although generally applicable, their use in plant pathology will be emphasized. We discuss, in some detail, strategies for the production and purification of these specific recombinant proteins. We give an overview of the literature and also provide some 'kitchen tricks', which are supported by some original data. Detailed protocols are included for the expression and purification of antibody fragments cloned into some of the frequently used vector systems. Thus, this chapter should provide a theoretical background as well as practical tips for expressing efficient diagnostic tools in *E. coli*. As a matter of definition, the term 'monoclonal antibody' will be used to refer to whole antibodies produced by animal cells, whereas the term 'recombinant antibody' will relate to bacterially-derived antibody fragments.

4.2 Potential reagents

Bacteria do not carry out glycosylation and intermolecular disulphide bond formation between heavy chains, two post-translational modifications necessary for the synthesis of complete immunoglobulin molecules. For this reason, recombinant antibody technology deals only with the production of antibody fragments. Recombinant antibodies that have been successfully produced in *E. coli* are shown in Figure 4.1. All these antibody fragments are expressed in the form of secretory proteins. The redox-conditions in the periplasmic space enable the appropriate folding and the correct formation of disulphide bonds to take place. Proteins may be harvested either from the periplasm or from culture supernatant. The first immunoglobulin fragments reported to be expressed in *E. coli* and purified in a functional form were Fv fragments (Skerra and Plückthun, 1988) and Fab fragments (Better *et al.*, 1988; Plückthun and Skerra, 1989) (Figure 4.1). Fv fragments consist of non-covalently linked V_H and V_L domains. Fab fragments are composed of the antibody light chain polypeptide and the heavy chain variable domain, followed by the C_{H1} domain. The light chain and the truncated heavy chain are covalently linked by a disulphide bond. Fab fragments have proven to be stable constructs and generally applicable to practical assays, whereas for Fv fragments associations between the V_H and V_L domains are often unstable (Cumber *et al.*, 1992). Fv have been stablilized by linking the individual V_H and

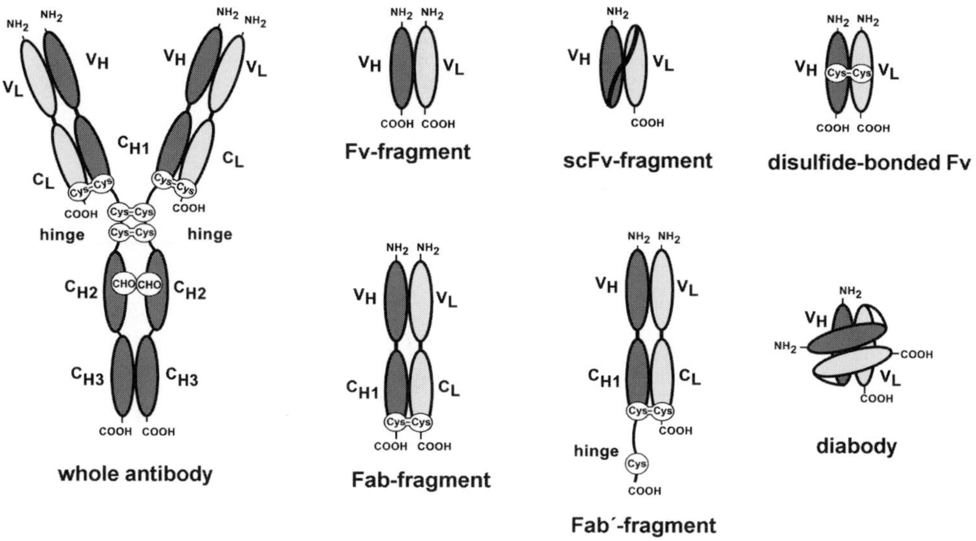

Figure 4.1 Schematic presentation of antibody fragments which may be expressed in *E. coli*, compared with a whole antibody. Intermolecular disulphide bonds (Cys–Cys) as well as the glycosylation sites (CHO) of the whole antibody are indicated (for further details see Section 4.2).

V_L domains by means of an artificial, flexible linker (Bird *et al.*, 1988; Huston *et al.*, 1988). The resulting fragment can thus be expressed as a single peptide chain, called a single-chain Fv-fragment (scFv) (Figure 4.1). Several other antibody fragment types have been cloned and successfully expressed in *E. coli.* These include Fab′ with a free cysteine residue for *in vitro* homodimerization (Carter *et al.*, 1992), disulphide-bonded Fv (Glockshuber *et al.*, 1990) and diabodies (Holliger *et al.*, 1993) (Figure 4.1). Also, smaller antibody fragments, consisting of only a single heavy chain variable region (Ward *et al.*, 1989) or even just of a modified portion of the V_H-domain (Bianchi *et al.*, 1994), have been reported to show antigen-binding activity. However, the most frequently used types of antibody fragments are scFv and Fab. These two kinds of proteins are well characterized and most published papers are based on either one of these constructs. Therefore, we will focus on these two types of antigen-binding proteins.

Antibody fragments are easily fused to additional proteins, protein domains or peptide-tags, so increasing enormously the variety of *E. coli*-derived antigen-binding proteins. Fusion proteins are constructed at gene level, by linking the sequence encoding for the fusion part in frame with the gene of the antibody fragment. Expression generates bi- or oligofunctional proteins, that often provide a means of detecting and purifying the antibody fragments, or combining them with novel effector functions. This enables broad applicability of recombinant antibodies in biochemistry and medicine. Figure 4.2 shows fusion proteins which can be used as immunodiagnostic reagents. In the case of Fab, the fusion parts are usually linked to the C-terminus of the V_H in order to avoid copurification of non-functional light chain homodimers (Constantine *et al.*, 1992). scFv fusion proteins are most commonly arranged to link the C-terminus of V_H to the N-terminus of V_L by a flexible linker-peptide and to connect the fusion partner to the C-terminus of V_L. However, other configurations, such as linkage of the N-terminus of V_H to the C-terminus of V_L to form a scFv (for example, Anand *et al.*, 1991; Pantoliano *et al.*, 1991) or connection of the fusion part to the N-terminus of a scFv (Kohl *et al.*, 1991), have also been reported.

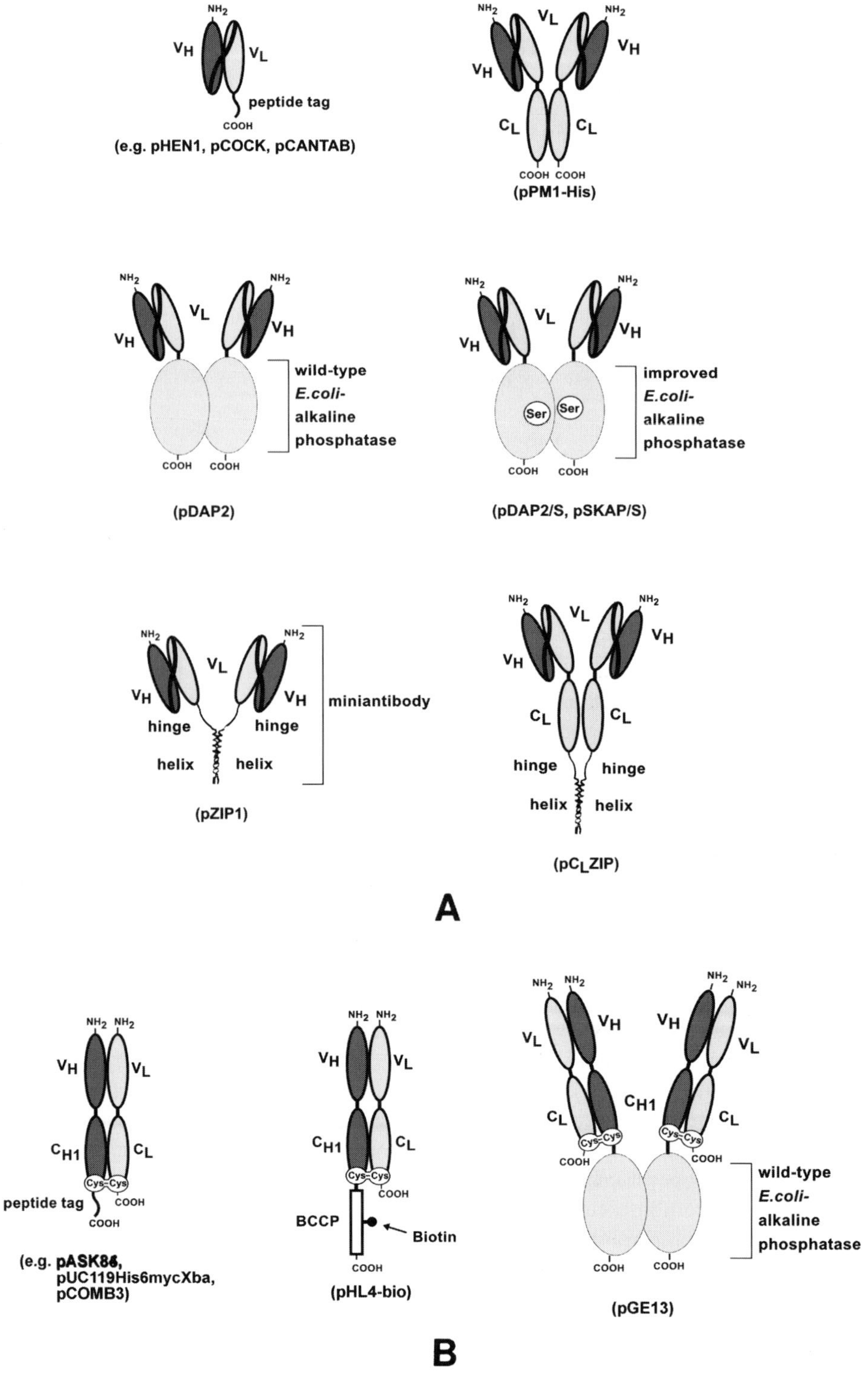

Figure 4.2 Schematic presentation of scFv fusion proteins (A) and Fab fusion proteins (B), suitable for immunoassays. Vectors for cloning and expressing the respective constructs are in parentheses and are described in Table 4.2 (for further details on fusion proteins see sections 4.2.1 and 4.2.2).

Table 4.1 Peptide tags for detection of antibody fragments

Peptide tag	Sequence	Detection with
c-myc (Ward *et al.*, 1989)	EQKLISEEDLN	Mab[1] 9E10 (Evan *et al.*, 1985)
E (Pharmacia, Sweden)	GAPVPYPDPLEPR	anti E-tag Mab (Pharmacia, Sweden)
FLAG (Power *et al.*, 1992)	DYKD(DDDK)[2]	Mabs M1 and M2 (Kodak, USA)
Histidine (Hochuli *et al.*, 1988)	HHHHH(H)[2]	Anti histidine tag Mab (von Strandmann *et al.*, 1995)
strep (Schmidt and Skerra, 1993)	SAWRHPQFGG	Streptavidin–enzyme conjugate

[1] Mab, monoclonal antibody.
[2] Amino acids in parentheses are omitted in some constructs.

4.2.1 ***Detecting reagents***

The immunological detection of an immobilized antigen by ELISA, or Western blotting, is conventionally achieved through the use of specific immunoglobulins, which are either chemically coupled to an enzyme, or are recognized by a secondary anti-immunoglobulin conjugate. Similarly, detection may be achieved by antibody fragments fused to peptide tags or protein domains, that are recognized by secondary conjugates, or by scFv– or Fab–enzyme fusion proteins, which can be detected directly.

Stretches of DNA encoding for short peptide sequences are often added to the end of an antibody fragment gene to produce antibody–peptide fusion proteins. These peptides allow the antibody fragment to be detected by means of a secondary antibody and are also useful for purifying the antibody fragment. The peptide tags most commonly used in recombinant antibody technology are described in Table 4.1. Several other peptides, such as the haemagglutinin tag (Thomann *et al.*, 1996) or the B tag (Wang *et al.*, 1996), have been fused to various proteins, but have not so far been routinely used in combination with recombinant antibodies. Detection of antibody fragments is also possible via antibody-constant domains present in Fab or scFv-constant domain fusion proteins. Constant domains may be detected by anti-Fab, anti-light chain or anti-gamma chain conjugates. These are inexpensive and widely available. Additionally, in the case of scFv-C_L fusion proteins (Figure 4.2A) dimers are formed, and binding is enhanced due to the increased avidity of the antibody (McGregor *et al.*, 1994). Another common approach to labelling of antigen-binding proteins is the attachment of biotin, which may be done by standard chemical procedures. Alternatively, biotinylation may be achieved by expression of antibody fragments fused to the C-terminal 101-amino acid polypeptide of the biotin carboxyl carrier protein (BCCP) subunit of *E. coli* acetyl-CoA carboxylase (Figure 4.2B). The BCCP subunit is biotinylated *in vivo*, on addition of d-biotin to the bacterial culture medium (Weiss *et al.*, 1994). This fusion protein may be detected using an avidin– or streptavidin–enzyme conjugate.

These approaches all require an additional step, in any immunoassay, to detect binding to the antigen by the recombinant antibody fragment. For routine applications it is preferable to minimize the number of working steps by employing antibody fragments covalently linked to an enzyme in order to reduce the time required for the assay and to reduce non-specific binding. Chemical coupling of scFv to enzymes is problematic, but the preparation of Fab conjugates by the maleimide method has been reported (Weiss and Van Regenmortel, 1989). To avoid the necessity of chemically coupling an antibody fragment to an enzyme molecule, antibody fragments have been genetically fused to *E. coli* alkaline phosphatase. Like chemically coupled conjugates, such fusion proteins combine both the binding properties of the antibody fragment and the enzymatic activity of the enzyme. The antigen can therefore be detected directly without the need for an additional incubation step. The genetically-derived antibody–enzyme fusion proteins are monosubstituted and produced reproducibly. In contrast, similar chemically-derived products may generate diverse multiple substituted conjugates which are often less effective in immunoassays (Daunert *et al.*, 1989). Furthermore, antibody fragments fused to *E. coli* alkaline phosphatase display an avidity effect, as the enzyme forms dimers (Figure 4.2). An scFv–*E. coli* alkaline phosphatase fusion protein was first described by Kohl *et al.* in 1991. Since then, several *E. coli* alkaline phosphatase scFv and Fab fusion proteins have been reported (Wels *et al.*, 1992; Ducancel *et al.*, 1993; Weiss and Orfanoudakis, 1994; Lindner *et al.*, 1997). A generally applicable system for rapid cloning and expression of scFv–alkaline phoshatase fusion proteins has been described (Kerschbaumer *et al.*, 1996), and used successfully in plant science (Harper *et al.*, 1997). However, these fusion proteins are less sensitive than chemically coupled immunoglobulins, due to the relatively low specific activity of *E. coli* alkaline phosphatase. To overcome this problem, an improved *E. coli* alkaline phosphatase was employed as a fusion partner for scFv (Kerschbaumer *et al.*, 1997). The substitution of an aspartate to a serine in the active site of the enzyme enhanced the specific activity 35-fold (Mandecki *et al.*, 1991) (Figure 4.2A). An scFv specific for *Beet necrotic yellow vein virus* (BNYVV, genus *Benyvirus*) fused to the mutated enzyme detected the plant pathogen with higher sensitivity than a well proven and reliable polyclonal conjugate, produced according to Koenig *et al.* (1984) (Kerschbaumer *et al.*, 1997). The vector for rapid cloning and expression of such an improved scFv–*E. coli* alkaline phosphatase fusion protein is shown in Figure 4.3A.

4.2.2 Coating reagents in ELISA

Many of the plant viruses of economic importance are present in their host plants at very low concentrations. In most cases, it is not possible to bind enough virus onto the ELISA plate by applying plant extract directly. Therefore, predominantly sandwich ELISA are performed in plant pathology, and the pathogen is trapped by a specific antibody coated onto the plate. Figure 4.4 depicts the different ELISA systems generally used for plant pathogen detection. In order to design an ELISA based completely on recombinant antibodies, it is necessary to produce a recombinant coating reagent. For efficient trapping of the antigen, the coating reagent has to be large enough to avoid conformational distortions of the antigen-binding region, which may occur when the protein binds to the plate. Moreover, it is preferable to use multimeric constructs to benefit from the avidity effect. It has been shown that a monomeric, non-fused scFv molecule does not trap virus, but that a miniantibody derived from the same scFv is an efficient coating reagent (Kerschbaumer *et al.*, 1997). The miniantibody employed in this study consisted of an scFv fused to the upper hinge region of mouse IgG3, in turn linked to a leucine zipper, a parallel coiled coil helix that forms very stable dimers (Pack and Plückthun, 1992) (Figure 4.2A). Higher coating efficiency is possible by enlarging the miniantibody and inserting a C_L-domain between the V_L of the scFv and the hinge region (Kerschbaumer, unpublished data) (Figure 4.2A). The corresponding vector for cloning and expression of such enlarged

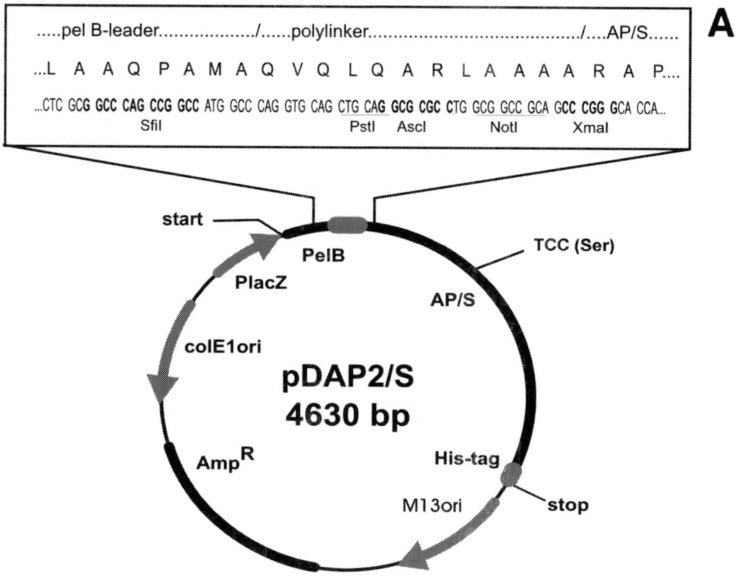

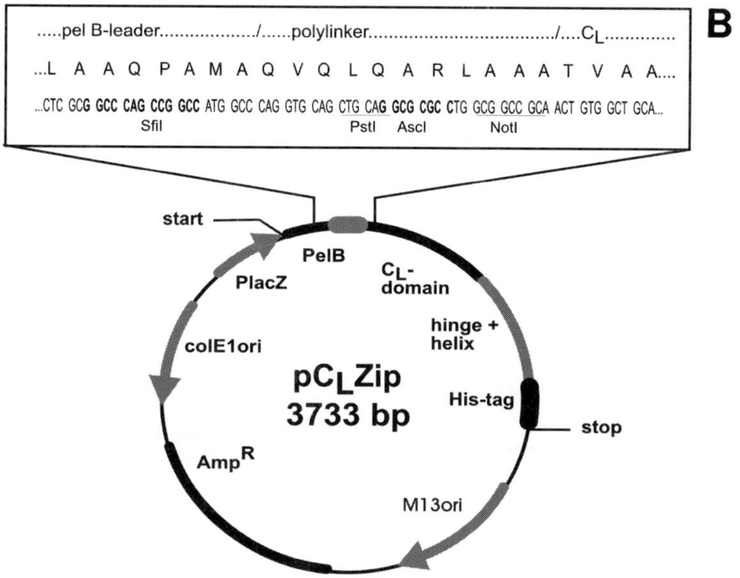

Figure 4.3 Maps and polylinker sequences of vectors pDAP2/S (A) and pC_LZIP (B). The unique restriction enzyme recognition sequences of the cloning sites are indicated (underlined/bold type). The pelB leader sequence (PelB) and the polylinker are located downstream of the start codon, followed either by the sequence coding for the C_L-domain, the hinge region and the amphipathic helix, or by the mutated ecphoAl-gene (AP/S). The sequences encoding the hexa-histidine-tag (His-tag) and the TAA stop codon complete the open reading frame in both cases. Also, the lac-promoter (PLacZ), the origin of replication (colE1ori), the M13 intergenic region (M13ori) and the β-lactamase-gene (Amp^R) are shown. pC_LZIP was constructed by inserting the coding sequence of the human IgG1 C_L-domain into pZIP1 (Kerschbaumer *et al.*, 1997). The combination of products from the vectors pDAP2/S and pZIP1 or pC_LZIP facilitates an ELISA based entirely on recombinant reagents. Figure 4.3A is reproduced from Kerschbaumer *et al.* (1997) with kind permission of Academic Press.

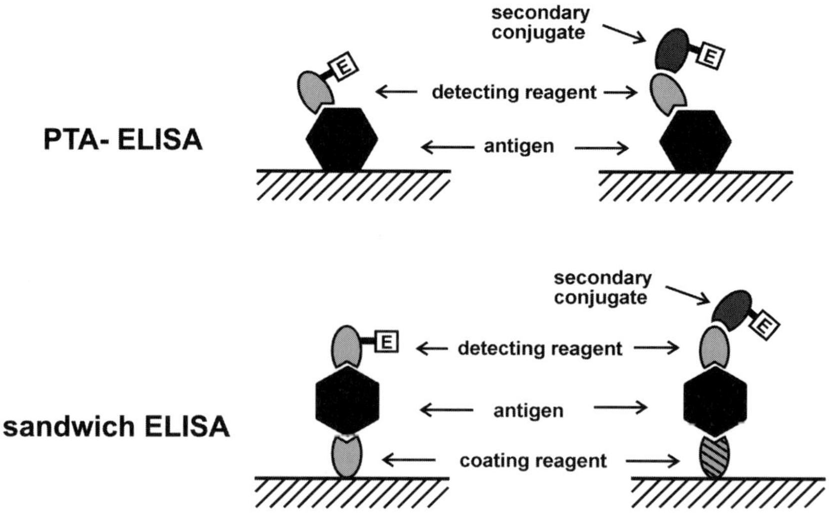

Figure 4.4 ELISA systems used for plant pathogen detection (Torrance, 1992). In a plate trapped antigen (PTA) ELISA, the plant pathogen is adsorbed directly to the plate (Ziegler *et al.*, 1995). This format is useful only at high antigen concentrations. In a sandwich ELISA, antigens are specifically trapped onto microtitre plates by a layer of a coating reagent. The plant pathogen can thus be immobilized even at low concentrations in a crude plant extract. In both formats the immobilized antigen can be bound by a detecting reagent that is either directly coupled to an enzyme (E) or is recognized by a secondary enzyme conjugate. This conjugate must not cross-react with the coating reagent.

miniantibodies is shown in Figure 4.3B. The inclusion of the leucine zipper results in a more stable dimerization of the protein compared with the antibody fragment–C_L fusion proteins without coiled coil helix. Di- and oligomerization of antibody fragments are reviewed by Plückthun and Pack (1997).

Recombinant coating reagents, made from antibody fragments derived from a hybridoma cell line, will have similar trapping properties as the parental monoclonal antibody. Some monoclonal antibodies are used successfully as coating reagents, whereas others show very poor trapping efficiency, although they have excellent binding properties in solution (Van Regenmortel, personal communication; Koenig, personal communication). Monoclonal antibodies which do not trap may not coat to the plate, or are inhibited by constituents of the plant extract. According to our experience only monoclonal antibodies with high trapping efficiency should be subcloned to design efficient recombinant coating reagents.

4.3 Expression

Efficient expression of heterologous proteins in *E. coli* depends on many factors (reviewed by Makrides, 1996). These include the elements of the expression vector, targeting and post-translational modifications of the protein, the strain of *E. coli* employed, as well as culture conditions. All these factors must be selected carefully, in order to express antibody fragments efficiently. The tendency of scFv and Fab to form insoluble inclusion bodies, their possible toxicity for the host cells and the instability of their genes often make production difficult. Despite these drawbacks, generally applicable vector systems and culture conditions have been described that allow expression of almost any antibody fragment in small laboratory scale as well as in fermenters.

4.3.1 *Commonly used expression vectors*

There are many vectors for cloning and expressing antibody fragments and fusion proteins thereof. The vectors are designed to provide convenient restriction sites, allowing the antibody gene to be inserted between a sequence coding for a leader peptide and the coding sequence for a fusion partner (which may be a peptide tag, enzyme or additional protein domain) being incorporated into the construct. The essential elements of vector constructs for expression of scFv and Fab fusion proteins are shown in Figure 4.5. Table 4.2 lists vectors for construction and expression of the potential diagnostic reagents which are displayed in Figure 4.2.

For antibody fragments to be produced as secretory proteins (see Section 4.2), the membrane transport mechanism of *E. coli* requires an efficient leader peptide at the N-terminus of the polypeptide chain. This signal sequence ensures translocation of the polypeptide across the inner membrane into the periplasmic space. Several leader sequences have been reported to be suitable. These peptides may be *E. coli*-derived, such as the ompA leader or the phoA leader (Skerra and Plückthun, 1991) or may have a different prokaryotic origin, such as the pelB leader (Better *et al.*, 1988). These are the three most commonly used signal sequences, but the bla leader (Skerra and Plückthun, 1991), ompF leader (Sawyer *et al.*, 1992) and stII leader (Carter *et al.*, 1992) sequences have also been used successfully. For the expression of Fab fragments, it is better to apply different leader sequences to the heavy and light chains to avoid recombination between homologous sequences. However, protein transport through the inner membrane also

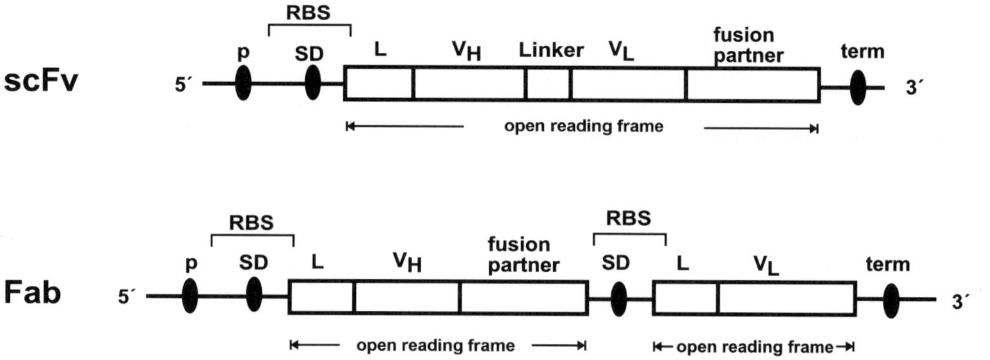

Figure 4.5 Vector elements for expression of scFv and Fab fusion proteins. The promoter (p) for regulation of transcription is followed by the ribosome binding site (RBS), with its Shine–Delgarno sequence (SD). The leader sequence (L) is located at the beginning of the coding sequence of the antibody fragment. For Fab expression, a leader sequence and RBS for each chain is necessary. Fusion partners may be a peptide tag, a protein or a protein domain. A transcription terminator (term) limits the monocistronic (for scFv) or dicistronic (for Fab) mRNA. Essential elements not indicated are the antibiotic resistance gene (for selection of plasmid harbouring bacteria) and the origin of replication (which determines plasmid copy number). For further details see Section 4.3.1.

Table 4.2 Vectors for cloning and expression of recombinant detection reagents

Vector	Construct	Promoter
pASK75 (Skerra, 1994b)	scFv→strep-tag	Tetracycline
pASK84 (Skerra, 1994a)	Fab→histidine-tag	lacUV5
pASK85 (Skerra, 1994b)	Fab→strep-tag	Tetracycline
pCANTAB 5E (Pharmacia, Sweden)	scFv→E-tag	lacZ
pC_LZIP (Kerschbaumer, unpublished)	scFv→C_L→hinge+helix→histidine-tag	lacZ
pCOCK (Engelhardt *et al.*, 1994)	scFv→c-myc-tag	lacZ
pCOMB3 (Barbas and Lerner, 1991)	Fab	lacZ
pDAP2 (Kerschbaumer *et al.*, 1996)	scFv→AP[1]→histidine-tag	lacZ
pDAP2/S (Kerschbaumer *et al.*, 1997)	scFv→AP/S[2]→histidine-tag	lacZ
pGE13 (Weiss and Orfanoudakis, 1994)	Fab→AP[1]	lacZ
pHEN1 (Hoogenboom *et al.*, 1991)	scFv→c-myc-tag	lacZ
pHFA (Lilley *et al.*, 1994)	scFv→FLAG-tag	lacZ
pHL4-bio (Weiss *et al.*, 1994)	Fab-BCCP	lacZ
pPM1-his (McGregor *et al.*, 1994)	scFv→C_L→histidine-tag	lacZ
pSKAP/S (Griep *et al.*, 1999)	scFv→AP/S[2]→histidine-tag	Tetracycline
pUC119His6mycXba (Griffiths *et al.*, 1994)	Fab→c-myc-tag→histidine-tag	lacZ
pZIP1 (Kerschbaumer *et al.*, 1997)	scFv→hinge+helix→histidine-tag	lacZ

All vectors mentioned harbour a β-lactamase gene mediating ampicillin resistance and an origin providing a high copynumber of plasmid.
[1] AP, wild-type *E. coli* alkaline phosphatase.
[2] AP/S, improved *E. coli* alkaline phosphatase (according to Mandecki *et al.*, 1991).

depends on several other elements and is a complex, not yet completely understood process (Pugsley, 1993). It is also possible to harvest insoluble aggregates of antibody fragments from cytoplasmic (for example, Buchner and Rudolph, 1991; McCartney *et al.*, 1991) or periplasmic inclusion bodies (for example, Denzin *et al.*, 1991; Pantoliano *et al.*, 1991). Although high levels of expression may be obtained, low yields of functional protein are often the final result of laborious refolding processes. Furthermore, refolding conditions must be carefully optimized for each protein sequence. For these reasons, we consider the production of antibody fragments as inclusion bodies to be of minor importance, and soluble expression under conditions avoiding formation of inclusion bodies to be the method of choice.

Expression of a heterologous protein is regulated by a promoter sequence that governs gene transcription. Promoters are under control of regulatory proteins, encoded either on the vector itself or on the *E. coli* genome (McClure, 1985; Lisser and Marglait, 1993; Goldstein and Doi, 1995). A useful promoter should be tightly controlled, ensuring minimal basal expression and undisturbed cell growth in the pre-induction phase. Methods for induction of mRNA formation, and hence of recombinant protein production, should be simple, efficient and cheap. To date, several promoters have been used for efficient expression of recombinant antibodies. Many cloning and expression vectors for antibody fragments contain the wild-type lacZ promoter. These include vectors derived from pUC119 (Vieira and Messing, 1987), such as pHEN1 and the pDAP vectors (see Table 4.2). The lacZ promoter is regulated by the lacI repressor, which is encoded on the bacterial genome. Repression occurs in the presence of glucose, and transcription can be induced by addition of isopropyl-β-thiogalactoside (IPTG) in the absence of glucose (De Bellis and Schwartz, 1990). A glucose-independent variant of this promoter, the lacUV5 promoter, has been used for Fab expression (Skerra, 1994a). Tighter, glucose-independent regulation is provided by vectors harbouring the tetracycline promoter and the respective repressor gene (Skerra, 1994b). Induction is achieved by addition of anhydro-tetracycline, a non-antibiotic derivative of tetracycline. Other promoters described for the expression of immunoglobulin fragments, such as the phoA promoter (Carter *et al.*, 1992) and the *E. coli* arabinose promoter (Clark *et al.*, 1997) are less commonly used. Besides a promoter for the initiation of mRNA synthesis, an expression vector also requires a transcription terminator, located downstream of the coding region to stop RNA elongation (Platt, 1986; Richardson and Greenblatt, 1996). Absence of a transcription terminator may substantially lower genetic stability and the level of protein production (Makrides, 1996, and citations therein). In the vectors most commonly used in antibody technology, efficient termination is provided by the M13 intergenic region, present in pUC119-derived vectors (Vieira and Messing, 1987; Smith, 1988), or by the lipoprotein transcription terminator inserted into many pASK vectors (Skerra, 1994a, 1994b).

Another important element of expression vectors is the ribosome binding site (RBS) with its Shine–Delgarno sequence. This region interacts with the 16S rRNA and is responsible for initiation of translation (Shine and Delgarno, 1974; Ringquist *et al.*, 1992; Voorma, 1996). In most cases the natural RBS of the respective leader sequence is employed for expression of antibody fragments (for example, Better *et al.*, 1988). Higher efficiency of the initiation of translation has been provided by two subsequent RBS. This design consists of a promoter and its natural RBS followed by the leader RBS and the open reading frame (Skerra, 1994a, 1994b).

4.3.2 Conditions for cell growth and protein expression

Conditions for growing *E. coli* harbouring antibody genes, and for expression of recombinant immunoglobulins are of immense importance and are often underestimated. Many problems accompanying recombinant antibody technology have been solved by optimizing growth parameters, such as culture medium and temperature. One of the major problems is the genetic instability of plasmids containing antibody genes. Cultures may be overgrown by cells which have lost their plasmids or which harbour a mutated antibody fragment gene. Cells not under the additional stress of recombinant expression obviously have an enormous advantage over bacteria with correct inserts, even at basal expression levels. Additionally, recombinant antibody genes are particularly sensitive to the bacterial recombination machinery. However, to ensure genetic stability of recombinant antibodies during the pre-induction phase of expression it is necessary to employ a vector that provides a tightly regulated promoter, as discussed in Section 4.3.1, but genetic stability cannot be guaranteed by a tight promoter alone. Interestingly, growth of bacteria at low temperature (25–30°C) dramatically increases the genetic stability and reliability of recombinant antibody clones. Figure 4.6 shows an experiment to emphasize the extraordinary importance of growth temperature even under non-inducing conditions. After induction, the burden on the cells increases enormously. The toxicity of the expressed antibody fragments for *E. coli* and the tendency of scFv and Fab to form inclusion bodies, or to overload the translocation machinery, may be a problem. Again, expression at low temperatures is a simple and efficient way to improve the yield of functional antibody fragments. This effect has been observed for a number of other recombinant proteins before (for example, Takagi *et al.*, 1988). Temperatures suitable for efficient expression of recombinant antibodies range from ambient temperature (Skerra and Plückthun, 1991) to 16°C (Kerschbaumer *et al.*, 1997). It has been shown that plasmid stability during the induction phase increases

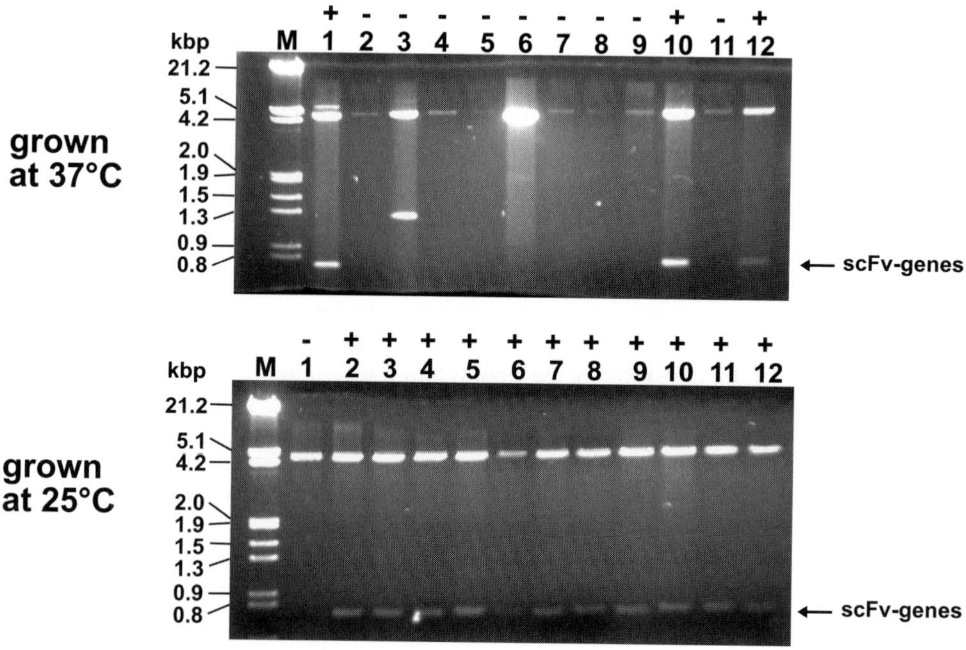

Figure 4.6 Stability of scFv genes at growth temperatures of 37°C and 25°C. *E. coli* TG-1 cells were infected with phage from an scFv phage-display library (Sheets *et al.*, 1998) and plated on two 2 × YT agar plates containing 2 per cent glucose and 100 μg/ml ampicillin (Sambrook *et al.*, 1989). One plate was grown overnight at 37°C, the other for 24 hours at 25°C. From each plate, 12 clones were selected at random and inoculated into 2 ml 2 × TY containing 100 μg ampicillin. Inoculants from the 37°C plate were incubated overnight, shaking at 37°C, and inoculants from the 25°C plate were incubated at 25°C for 24 hours. All cultures had a similar cell-density, and plasmid DNA was prepared by the pLASmix-miniprep kit (Talent, Italy) according to the manufacturer's instructions. Plasmid DNA was incubated with the restriction enzymes *Sfi* I and *Not* I, which have unique restriction sites at the 5′-end and 3′-end of the scFv gene, respectively. Analysis of the digested DNA on a 1 per cent agarose gel revealed only three of the clones grown at 37°C to have an insert of the correct size (750 bp). The remaining clones had either lost their scFv genes or contained a modified insert. Also, the plasmid yields varied, demonstrating the different extends of plasmid loss. On the other hand, almost all the clones grown at 25°C (except clone 1) had inserts of the correct size and gave comparable yields of DNA. Presence of an scFv gene was verified by PCR amplification of the plasmid DNA templates. Clones resulting in PCR products of the correct size (+) and those resulting in no amplified products or products of the wrong size (–) are indicated. These results emphasize the importance of growth conditions for the genetic stability of antibody genes (even when not being expressed), and that genetic stability can be provided by growth at lower temperatures. M, marker (*EcoR* I and *Hind* III digested DNA of phage lambda); kbp, kilobase pairs.

enormously if the temperature is reduced to 16°C (Kerschbaumer *et al.*, 1997). The formation of inclusion bodies is also minimized at low temperatures (Cabilly, 1989; Schein, 1993). This is probably because the rates of biochemical reactions are substantially decreased at low temperatures, thus providing sufficient time for proper folding of the antibody fragment (Oppenheim *et al.*, 1996).

Besides temperature, the presence of carbohydrate, which serves as an efficient energy source, is important for increasing yield. Glucose, for example, should be present after induction in the case of glucose-independent promoters. However, many of the vectors commonly used employ the lacZ promoter, which is repressed by glucose. Here, glucose is necessary in the pre-induction phase, but must be removed for efficient expression and replaced with another carbon source such as glycerol (Kerschbaumer *et al.*, 1996). Figure 4.7A demonstrates the importance of choosing the best expression conditions to produce a detection reagent, in this case, specific for BNYVV.

Additionally, protein production is influenced by the bacterial strain. Several different *E. coli* strains have been used successfully in the expression of antibody fragments, for example, TG-1 (McCafferty *et al.*, 1990), HB2151 (Marks *et al.*, 1991), XL1-Blue (Barbas and Lerner, 1991), RV308 (Pack *et al.*, 1993) and JM83 (Skerra, 1994a, 1994b), but no systematic comparison of these strains has been reported. Suppressor strains such as TG-1 are used in combination with phage display vectors such as pHEN1 (Hoogenboom *et al.*, 1991) or pCOCK (Engelhardt *et al.*, 1994) to produce phage particles displaying antibody on their surface (McCafferty *et al.*, 1990). These vectors separate the coding region of the antibody fragment and the fd gene3 with an amber stop codon. This is not recognized as a stop codon in suppressor strains, and instead, a glutamate is synthesized. For soluble expression of the antibody fragments these plasmids are transformed into non-suppressor strains, for example, HB2151 (Marks *et al.*, 1991). However, TG-1 cells are often used for expressing soluble protein after subcloning antibodies into vectors providing an ochre or opal stop codon (Griffiths *et al.*, 1994; Schier *et al.*, 1996).

The most commonly used vectors for production of antibody fragments harbour the ampicillin resistance gene and the lacZ promoter (see Table 4.2). Section 4.5 provides an optimized protocol for laboratory-scale production of scFv and Fab fusion proteins in shaker flasks using such vectors. We must emphasize that there are considerable differences between antibody fragments regarding their genetic stability or toxicity to the host cell. Thus, yields differ, depending on the scFv or Fab sequence, and each fusion partner.

4.3.3 Strategies for scaling up production

Large-scale production of antibody fragments is usually achieved in fermenters. It is possible to monitor growth conditions continually during fermentation, allowing a constant pH and pO_2 to be maintained. Batch fermentations may result in a five- to tenfold higher yield than is possible in shaker flasks (Figure 4.7B). Such batch fermentations yield comparably low cell densities, therefore in most cases fed-batch procedures are performed for the production of recombinant antibodies in fermenters (Carter *et al.*, 1992; Pack *et al.*, 1993; Horn *et al.*, 1996). Fed-batch fermentations allow growth at low but constant carbohydrate concentrations by controlled addition of carbohydrate during cell growth, thus minimizing formation of toxic side products (for example, acetate) and resulting in much higher cell densities and improved expression yields (reviewed by Lee, 1996). Fed-batch fermentation of antibody fragments has been reported to reach cell densities of up to 145 g dry biomass per litre and 3 g of functional antibody fragment per litre (Horn *et al.*, 1996).

The yield for a given antibody fragment may be limited when the vector systems mentioned in Table 4.2 are used for expression, as these vectors harbour an ampicillin resistance gene coding for β-lactamase. The major drawback of using β-lactamase is that this enzyme is secreted into the culture supernatant, thus degrading the antibiotic. Within

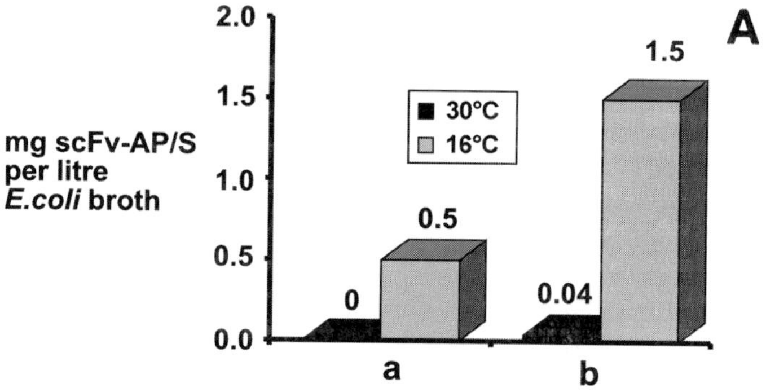

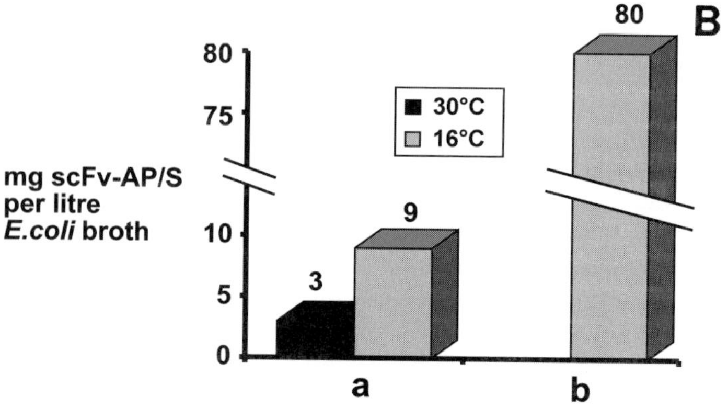

Figure 4.7 Comparison of yields for various expression conditions. In each experiment, an anti-BNYVV scFv, fused to an improved *E. coli* alkaline phosphatase (AP/S), was expressed by means of pDAP/2S (Kerschbaumer *et al.*, 1997) in *E. coli* TG-1 cells. Amount of product was determined as described (Kerschbaumer *et al.*, 1997). In all cases more than 95 per cent of the product was found in the supernatant. (A) Influence of medium and temperature. (a) Production with no added carbohydrate during induction phase. Expression was performed as described in Section 4.5.1, except for the omission of glycerol in the induction phase. After induction by addition of IPTG, cells were grown with shaking at 30°C or 16°C. (b) Production with carbohydrate during induction phase. Expression was performed as described in Section 4.5.1. After induction by addition of IPTG, cells were grown with shaking at 30°C or 16°C. (B) Influence of resistance gene and production in fermenter. (a) Production with tetracycline resistance in shaker flask. pDAP2/S harbouring the scFv gene was modified by inserting the tetracycline resistance gene of pBR322 into the unique *Xmn* I site. Cells harbouring this modified vector were grown as described in Section 4.5.1, except that the medium contained 15 μg/ml tetracycline instead of ampicillin. After induction by addition of IPTG, cells were grown at 30°C or 16°C, respectively. The product yield reached a maximum after five days. (b) Production with tetracycline resistance in fermenter (batch fermentation). Cells harbouring the tetracycline resistant version of pDAP2/S containing the respective scFv gene (see above) were grown at 28°C in a 1 litre fermenter (Chemap, Switzerland) in synthetic M9-medium (Sambrook *et al.*, 1989) supplemented with 12 g glucose, 60 mg thiamine and 15 mg tetracycline. When the culture reached an OD_{600} of 10 the glucose was used up. The culture was then cooled down to 16°C, and 100 g of glycerol and IPTG (to a final concentration of 1 mM) were added. During the whole process an oxygen saturation of 50 per cent and a pH of 6.8 was kept constant. Cells were grown to a density of 7 g bacteria dry mass per litre, and the product yield reached a maximum after six days.

a very short time the selection pressure is destroyed and bacteria which have lost their plasmid can overgrow the culture (Kerschbaumer *et al.*, 1997). To avoid plasmid loss, other resistance genes such as the kanamycin resistance gene (Pack *et al.*, 1993) or the tetracycline resistance gene (Kerschbaumer, unpublished; Figure 4.7B) have been used as the selection marker. The products of these resistance genes are not secreted, and the selection pressure is maintained throughout the cultivation. As a consequence, yield increases, since during the entire induction phase only bacteria harbouring the recombinant plasmid are able to grow. Figure 4.7 demonstrates how the yield may be increased in shaker flask experiments, by replacing ampicillin resistance (Figure 4.7A) with tetracycline resistance (Figure 4.7B).

Other strategies for improvement of antibody fragment expression have been investigated. Framework mutations (Forsberg *et al.*, 1997; Nieba *et al.*, 1997) or alternative linker sequences (Turner *et al.*, 1997) have been reported to improve expression of scFv. Expression in thioredoxin reductase-deficient *E. coli* strains has been described, where disulphide bond formation occurs in the cytoplasm (Proba *et al.*, 1995), thus making secretion of the antibody fragments unnecessary. Co-expression of human protein disulphide isomerase has been shown to increase the yield of an antibody Fab′ fragment expressed in *E. coli* (Humphreys *et al.*, 1996). Co-expression of molecular chaperones to mediate efficient post-translational folding is also possible (Wall and Plückthun, 1995). Furthermore, alternative promoters or systematically optimized RBS (Simmons and Yansura, 1996) have led to significant increases in the expression of various other heterologous proteins, and these approaches may be applied to antibody fragments. Optimization does not necessarily mean finding the strongest promoter or RBS. Reducing the rate of protein synthesis disfavours protein misfolding and improves cell viability, and thus often results in larger amounts of soluble product (Hockney, 1994; Georgiou and Valax, 1996). Although these strategies are currently not used routinely in recombinant antibody technology, they may play an important role in the future.

To obtain high expression levels of antibody fragments, hosts other than *E. coli* have been used. These include other prokaryotes such as *Bacillus subtilis* (Wu *et al.*, 1993), filamentous fungus (Nyyssönen *et al.*, 1993), yeast (for example, Bowdish *et al.*, 1991; Ridder *et al.*, 1995), insect cells (Ward *et al.*, 1995; Kretzschmar *et al.*, 1996), plant cells and plants (Firek *et al.*, 1993; Schouten *et al.*, 1997). Among these possibilities, we consider production by the yeast *Saccharomyces cerevisiae* (for example, Bowdish *et al.*, 1991, Edqvist *et al.*, 1991) or *Pichia pastoris* (for example, Luo *et al.*, 1995; Ridder *et al.*, 1995; Eldin *et al.*, 1997) to be the most promising options to substitute production in *E. coli*. Yeast combines the efficiency of the eukaryotic secretion machinery with fast growth, without the requirement for complex culture media. However, the necessary cloning is more laborious and time consuming than for bacteria.

4.4 Purification

When expressed in bacterial cells, recombinant antibody fragments are usually harvested from the periplasm or from the culture supernatant. Localization is dependent on several variable factors, such as temperature during expression, duration of the induction phase and choice of production method (for example, in shaker flasks or by fermentation). Also, the antibody sequence itself has a strong influence on the leakiness of the outer membrane. The expression protocol given in Section 4.5 usually results in secretion of the protein into the supernatant, but a certain amount of product can also be found in the periplasm. The purification protocols include preparation of periplasmic extracts and concentrated culture supernatants, each followed by an affinity chromatography step to remove *E. coli* proteins. Protease inhibitors are omitted in most protocols described for harvesting antibody fragments from the periplasm or supernatant, because most of the *E. coli* proteases are not secreted and thus proteolytic activity in the periplasm is low.

4.4.1 Preparation of periplasmic extracts and concentrated culture supernatants

In order to harvest the contents of the periplasm, cells are pelleted and resuspended in an extraction buffer to release protein from the periplasmic space. These buffers usually contain EDTA to weaken the outer membrane, and high concentrations of NaCl or sucrose for hyperosmotic disruption of the outer membrane (Lindner *et al.*, 1992; Skerra, 1994a, 1994b). Spheroplasts are then separated from the soluble periplasmic protein by centrifugation. Depending on the affinity chromatography technique being used, the extraction buffer may require to be exchanged, for example by dialysis, before continuing with the purification procedure.

Purification from the culture medium is in general more laborious, but often yields considerably more of the desired protein than from the periplasm. After clearing the broth by centrifugation, concentration and/or dialysis is usually necessary. The easiest way of concentrating the antibody fragment is by ammonium sulphate precipitation (Kerschbaumer *et al.*, 1996) or by ultrafiltration (Marks *et al.*, 1991).

4.4.2 Affinity chromatography

Whole antibodies are usually purified by affinity chromatography on protein A or protein G columns (Jungbauer *et al.*, 1989). In the case of antibody fragments these methods are unsuitable, because of the absence of constant regions. Only Fab fragments can be purified by means of protein G because of the presence of the C_{H1}-domain (Carter *et al.*, 1992). Those Fab or scFv containing a kappa light chain can be purified via protein L from *Peptostreptococcus magnus* (Nilson *et al.*, 1993). This affinity ligand binds immunoglobulins primarily through the variable domain of kappa light chains, without interfering with antigen-binding sites. Antibody fragments containing a lambda light chain are not suitable for protein L-based purification.

Affinity purification of recombinant antibodies is most easily achieved by means of a peptide tag (Table 4.1). Many of these tags are themselves recognized by a monoclonal antibody, for example the FLAG tag (Knappik and Plückthun, 1994), and so allow affinity purification of the antibody fragment–peptide fusion protein. However, monoclonal antibodies are often expensive and elution conditions are often harsh. Affinity tags, other than those based on antibody–antigen interactions, have been described. These include the strep tag (Schmidt and Skerra, 1994), the histidine tag (Hochuli *et al.*, 1988), maltose-binding protein (Malinowski *et al.*, 1992) and calmodulin (Neri *et al.*, 1995).

Recombinant antibodies fused to a histidine tag may be purified by immobilized metal affinity chromatography (IMAC). This procedure is based on the affinity of several consecutive histidine residues to metal ions. Recombinant proteins with a histidine tag will bind to a chelating gel matrix charged with metal ions (most commonly used: Ni^{2+} and Zn^{2+}). The application of this method to antibody fragment purification was described and discussed in detail by Lindner *et al.* (1992). In the case of purification from supernatant, the small peptides present in a complex broth must first be removed by dialysis before continuing with IMAC and, depending on the antibody construct to be purified, concentration of the culture media may be necessary. In some cases, when the affinity of the product to the chelating gel is poor (perhaps because the tag is not completely accessible) and expression was performed in complex media, it might be favourable to do an initial gel filtration chromatography step. Most of the vectors discussed (see Table 4.2) encode for a C-terminal histidine tag, and so an IMAC protocol is described in Section 4.5.

4.5 Protocols

The most frequently used vector systems for the production of antigen-binding proteins in *E. coli* harbour the ampicillin resistance gene, regulate transcription by the lacZ promoter and provide a C-terminal histidine tag (see Table 4.2). Expression and purification may be performed according to the following protocols if such vectors are employed. These protocols are excerpts from Kerschbaumer *et al.* (1997) with some alterations and are reproduced with kind permission of Academic Press.

4.5.1 *Expression protocol*

1. *E. coli* TG-1 cells containing the plasmid construct are incubated overnight at 28°C in M9ZB medium containing 1 per cent glucose and 100 μg/ml ampicillin.
2. The culture is diluted 1:200 into fresh M9ZB medium containing 1 per cent glucose and 100 μg/ml ampicillin, and grown at 28°C to an OD_{600} of 0.5.
3. The culture is centrifuged (4500 **g** for five minutes at room temperature) and the cells are resuspended in fresh M9ZB containing 1 per cent glycerol and 100 μg/ml ampicillin. IPTG is added immediately to a final concentration of 1 mM.

The centrifugation step can be omitted, if the overnight culture is diluted in M9ZB containing 0.1 per cent glucose and 100 μg/ml ampicillin (step 2). In this case, glycerol and IPTG are added at an OD_{600} of 0.5 to the same final concentrations as above. This procedure is easier, especially when several cultures are being examined, but results in a lower yield.

4. The culture is incubated at 16°C for 15 hours (for harvesting protein from the periplasm) or for 48–72 hours (supernatant expression).

M9ZB medium

Solution A:	10 g bacto tryptone, 5 g bacto yeast extract, 5 g NaCl and 1 g NH_4Cl dissolved in 800 ml distilled water;
Solution B:	3 g KH_2PO_4 and 6 g Na_2HPO_4 dissolved in 100 ml distilled water;
Solution C:	1 M aqueous $MgSO_4$ solution;
Solution D:	10 g glucose in 100 ml water,
or	1 g glucose in 100 ml water,
or	10 g glycerol in 100 ml water.

Solutions A, B, C and D are prepared and autoclaved separately. Then 800 ml solution A, 100 ml solution B, 1 ml solution C and 100 ml of a solution D are mixed to give one litre M9ZB-medium.

4.5.2 Preparation of the periplasmic extract and the supernatant

Periplasmic extract

Cells are harvested by centrifugation (10 000 **g** at 4°C for 10 minutes) and resuspended in 200 mM sodium borate buffer, pH 8.0, containing 1 M NaCl and 1 mM EDTA (50 ml ice cold buffer per 1 litre of *E. coli* culture broth), and placed on ice for 30 minutes to release the fusion protein from the periplasmic space. After centrifugation (18 000 **g** at 4°C for 20 minutes) the supernatant is recovered and the pH adjusted to 7.3 by adding 1 M KH_2PO_4. In order to prepare the periplasmic extract for the subsequent IMAC purification, phosphate buffered saline (PBS; 10 mM potassium phosphate, 140 mM NaCl, pH 7.3) containing 1 M imidazole is added to give a final concentration of 10 mM imidazole.

Supernatant

The culture supernatant is obtained by centrifugation (10 000 **g** at 4°C for 10 minutes).

To prepare the supernatant for IMAC purification, one of the following approaches can be chosen according to the construct and its affinity to the Zn^{2+}-charged gel:

1. Ammonium sulphate (470 g) is added to 1 litre of supernatant and stirred on ice for 1 hour to precipitate protein (depending on the construct to be purified, less ammonium sulphate may be sufficient). The precipitate is pelleted at 18 000 **g** for 20 minutes and redissolved in 50 ml PBS. The concentrate is dialysed against PBS. Finally, solid NaCl and PBS containing 1 M imidazole are added to give final concentrations of 1 M NaCl and 10 mM imidazole.
2. The supernatant is dialysed directly against PBS. Solid NaCl and PBS containing 1 M imidazole are added to give final concentrations of 1 M NaCl and 10 mM imidazole.

For an uncharacterized construct, we recommend starting with protocol A.

4.5.3 Purification by immobilized metal affinity chromatography (IMAC)

1. A column containing 5 ml iminodiacetic gel matrix is rinsed with 25 ml of water.
2. The gel is charged with 25 ml of a 0.1 M aqueous $ZnCl_2$ solution and then rinsed again with 25 ml of water.
3. The gel is equilibrated with 25 ml PBS containing 1 M NaCl (PBS/1 M NaCl).
4. The sample is prepared as described in Section 4.5.2 and insoluble aggregates are removed by centrifugation (18 000 **g**, 10 minutes). The clarified sample is loaded onto the column.
5. After loading, the gel is washed with 25 ml PBS/1 M NaCl/10 mM imidazole and then with 25 ml PBS/1 M NaCl/20 mM imidazole.
6. Elution is performed using a step gradient: 5 ml PBS/1 M NaCl/40 mM imidazole, 5 ml PBS/1 M NaCl/60 mM imidazole, 5 ml PBS/1 M NaCl/80 mM imidazole, 5 ml

PBS/1 M NaCl/120 mM imidazole and finally 10 ml of PBS/1 M NaCl/50 mM EDTA. 1 ml fractions are collected and subjected to SDS-PAGE and silver staining.

Before reuse, the gel should be stripped of any residual material by washing with an additional 25 ml PBS/1 M NaCl/50 mM EDTA. It is then rinsed with water and re-charged as described above.

Acknowledgements

We thank Sonja Hirschl and Karin Bartonek for their technical assistance to obtain the original data as well as Dr Karl Bayer for support in fermentation.

The authors are indebted to the European Union (AIR3 project CT94-1046) for financial support.

References

ANAND, N.N., MANDAL, S., MACKENZIE, C.R., SADOWSKA, J., SIGURSKJOLD, B., YOUNG, N.M., *et al.*, 1991, Bacterial expression and secretion of various single-chain Fv genes encoding proteins specific for a salmonella serotype B O-antigen, *Journal of Biological Chemistry*, **266**, 21874–21879.

BARBAS III, C.F. and LERNER, R.A., 1991, Combinatorial immunoglobulin libraries on the surface of phage (Phabs): rapid selection of antigen-specific Fabs, *Methods: A Companion to Methods in Enzymology*, **2**, 119–124.

BETTER, M., CHANG, C.P., ROBINSON, R.R. and HORWITZ, A.H., 1988, *Escherichia coli* secretion of an active chimeric antibody fragment, *Science*, **240**, 1041–1043.

BIANCHI, E., VENTURINI, S., PESSI, A., TRAMONTANO, A. and SOLLAZZO, M., 1994, High level expression and rational mutagenesis of a designed protein, the minibody, *Journal of Molecular Biology*, **236**, 649–659.

BIBILA, T.A. and ROBINSON, D.K., 1995, In pursuit of the optimal fed-batch process for monoclonal antibody production, *Biotechnology Progress*, **11**, 1–13.

BIRD, R.E., HARDMANN, K.D., JACOBSON, J.W., JOHNSON, S., KAUFMAN, B.M., LEE, S.M., *et al.*, 1988, Single-chain antigen-binding proteins, *Science*, **242**, 423–426.

BOWDISH, K., TANG, Y., HICKS, J.B. and HILVERT, D., 1991, Yeast expression of a catalytic antibody with chorismate mutase activity, *Journal of Biological Chemistry*, **266**, 11901–11908.

BUCHNER, J. and RUDOLPH, R., 1991, Renaturation, purification and characterization of recombinant Fab-fragments produced in *Escherichia coli*, *Biotechnology*, **9**, 157–162.

CABILLY, S., 1989, Growth at sub-optimal temperatures allows the production of functional, antigen-binding Fab fragments in *Escherichia coli*, *Gene*, **85**, 553–557.

CARTER, P., KELLEY, R.F., RODRIGUES, M.L., SNEDECOR, B., COVARRUBIAS, M., VELLIGAN, M.D., *et al.*, 1992, High level *Escherichia coli* expression and production of a bivalent humanized antibody fragment, *Biotechnology*, **10**, 163–167.

CLARK, M.A., HAMMOND, F.R., PAPAIOANNOU, A., HAWKINS, N.J. and WARD, R.L., 1997, Regulation and expression of human Fabs under the control of the *Escherichia coli* arabinose promoter, P_{BAD}, *Immunotechnology*, **3**, 217–226.

CONSTANTINE, K.L., GOLDFARB, V., WITTEKIND, M., ANTHONY, J., NG, S.-C. and MUELLER, L., 1992, Sequential ^{1}H and ^{15}N NMR assignments and secondary structure of a recombinant anti-digoxin antibody V_L domain, *Biochemistry*, **31**, 5033–5043.

CUMBER, A.J., WARD, E.S., WINTER, G., PARNELL, G.D. and WAWRZYNCZAK, E.J., 1992, Comparative stabilities of a recombinant mouse antibody FvCys fragment and a bis FvCys conjugate, *Journal of Immunology*, **149**, 120–126.

DAUNERT, S., PAYNE, B.R. and BACHAS, L.G., 1989, Pyruvate carboxylase as a model for oligosubstituted enzyme-ligand conjugates in homogenous enzyme immunoassays, *Analytical Chemistry*, **61**, 2160–2164.

De Bellis, D. and Schwartz, I., 1990, Regulated expression of foreign genes fused to *lac*: control by glucose levels in growth medium, *Nucleic Acids Research*, **18**, 1311.

Denzin, L.K., Whitlow, M. and Voss, E.W. Jr, 1991, Single-chain site-specific mutations of fluorescein-amino acid contact residues in high affinity monoclonal antibody 4–4–20, *Journal of Biological Chemistry*, **266**, 14095–14103.

Ducancel, F., Gillet, D., Carrier, A., Ljeunesse, E., Menez, A. and Boulain, J.C., 1993, Recombinant colorimetric antibodies: construction and characterization of a bifunctional F(ab)2/alkaline phosphatase conjugate produced in *Escherichia coli, Biotechnology (NY)*, **11**, 601–605.

Edqvist, J., Keranen, S., Penttila, M., Straby, K.B. and Knowles, J.K., 1991, Production of functional IgM Fab fragments by *Saccharomyces cerevisiae*, *Journal of Biotechnology*, **20**, 291–300.

Eldin, P., Pauza, M.E., Hieda, Y., Lin, G., Murtaugh, M.P., Pentel, P.R. and Pennell, C.A., 1997, High-level secretion of two antibody single chain Fv fragments by *Pichia pastoris*, *Journal of Immunological Methods*, **201**, 67–75.

Engelhardt, O., Grabherr, R., Himmler, G. and Rüker, F., 1994, Two-step cloning of antibody variable domains in a phage display vector, *Biotechniques*, **17**, 44–46.

Evan, G.I., Lewis, G.K., Ramsey, G. and Bishop, J.M., 1985, Isolation of monoclonal antibodies specific for human c-myc proto-oncogene product, *Molecular and Cellular Biology*, **5**, 3610–3616.

Firek, S., Draper, J., Owen, M.R., Gandecha, A., Cockburn, B. and Whitelam, G.C., 1993, Secretion of a functional single-chain Fv protein in transgenic plants and cell suspension cultures, *Plant Molecular Biology*, **23**, 861–870.

Forsberg, G., Forsgren, M., Jaki, M., Norin, M., Sterky, C., Enhörning, Å., *et al.*, 1997, Identification of framework residues in a secreted recombinant antibody fragment that control production level and localization in *Escherichia coli*, *Journal of Biological Chemistry*, **272**, 12430–12436.

Georgiou, G. and Valax, P., 1996, Expression of correctly folded proteins in *Escherichia coli, Current Opinion in Biotechnology*, **7**, 190–197.

Glockshuber, R., Malia, M., Pfitzinger, I. and Plückthun, A., 1990, A comparison of strategies to stabilize immunoglobulin Fv-fragments, *Biochemistry*, **29**, 1362–1367.

Goldstein, M.A. and Doi, R.H., 1995, Procaryotic promoters in biotechnology, *Biotechnology Annual Review*, **1**, 105–128.

Griep, R.A., van Twisk, C., Kerschbaumer, R.J., Harper, K., Torrance, L., Himmler, G., *et al.*, 1999, pSKAP/S: an expression vector for the production of single-chain Fv alkaline phosphatase fusion-proteins, Genbank Accession no. AF083070.

Griffiths, A., Williams, S., Hartley, O., Tomlinson, I., Waterhouse, P., Crosby, W., *et al.*, 1994, Isolation of high affinity human antibodies directly from large synthetic repertoires, *EMBO Journal*, **13**, 3245–3260.

Harper, K., Kerschbaumer, R.J., Ziegler, A., Macintosh, S.M., Cowan, G.H., Himmler, G., *et al.*, 1997, A scFv-alkaline phosphatase fusion protein which detects potato leafroll luteovirus in plant extracts by ELISA, *Journal of Virological Methods*, **63**, 237–242.

Hochuli, E., Bannwarth, W., Döberli, H., Gentz, R. and Stüber, D., 1988, Genetic approach to facilitate purification of recombinant proteins with a novel metal chelate absorbent, *Biotechnology*, **6**, 1321–1325.

Hockney, R.C., 1994, Recent developments in heterologous protein production in *Escherichia coli*, *Trends in Biotechnology*, **12**, 456–463.

Holliger, P., Prospero, T. and Winter, G., 1993, 'Diabodies': small bivalent and bispecific antibody fragments, *Proceedings of the National Academy of Sciences of the USA*, **90**, 6444–6448.

Hoogenboom, H.R., Griffiths, A.D., Johnson, K.S., Chiswell, D.J., Hudson, P. and Winter, G., 1991, Multi-subunit proteins on the suface of filamentous phage: methodologies for displaying antibody (Fab) heavy and light chains, *Nucleic Acids Research*, **19**, 4133–4137.

HORN, U., STRITTMATTER, W., KREBBER, A., KNÜPFER, U., KUJAU, M., WENDEROTH, R., *et al.*, 1996, High volumetric yields of functional dimeric miniantibodies in *Escherichia coli*, using an optimized expression vector and high-cell-density fermentation under non-limited growth conditions, 1996, *Applied Microbiology and Biotechnology*, **46**, 524–532.

HUMPHREYS, D.P., WEIR, N., LAWSON, A., MOUNTAIN, A. and LUND, P.A., 1996, Co-expression of human protein disulphide isomerase (PDI) can increase the yield of an antibody Fab′ fragment expressed in *Escherichia coli*, *FEBS Letters*, **380**, 194–197.

HUSTON, J.S., LEVINSON, D., MUDGETT-HUNTER, M., TAI, M.S., NOVOTNY, J., MARGOLIES, M.N., *et al.*, 1988, Protein engineering of antibody binding sites: recovery of specific activity in an anti-digoxin single chain Fv analogue produced in *Escherichia coli*, *Proceedings of the National Academy of Sciences of the USA*, **86**, 5879–5883.

JUNGBAUER, A., TAUER, C., REITER, M., PURTSCHER, M., WENISCH, E., STEINDL, F., *et al.*, 1989, Comparison of protein A, protein G and copolymerized hydroxyapatite for the purification of human monoclonal antibodies, *Journal of Chromatography*, **476**, 257–268.

KERSCHBAUMER, R.J., HIRSCHL, S., IBL, M., SCHWAGER, C. and HIMMLER, G., 1996, pDAP2: a vector for construction of alkaline phosphatase fusion-proteins, *Immunotechnology*, **2**, 145–150.

KERSCHBAUMER, R.J., HIRSCHL, S., KAUFMANN, A., IBL, M., KOENIG, R. and HIMMLER, G., 1997, Single-chain Fv fusion proteins suitable as coating and detecting reagents in a double antibody sandwich enzyme linked immunosorbent assay, *Analytical Biochemistry*, **249**, 219–227.

KNAPPIK, A. and PLÜCKTHUN, A., 1994, An improved affinity tag based on the FLAG-peptide for the detection and purification of recombinant antibody fragments, *Biotechniques*, **17**, 754–761.

KOENIG, R., LESEMANN, D.-E. and BURGERMEISTER, W., 1984, Beet necrotic yellow vein virus: purification, preparation of antisera and detection by means of ELISA, immunosorbent electronmicroscopy and electro-blot immunoassay, *Phytopathologische Zeitschrift (Journal of Phytopathology)*, **111**, 244–250.

KOHL, J., RÜKER, F., HIMMLER, G., RAZAZZI, E. and KATINGER, H., 1991, Cloning and expression of an HIV-1 specific single-chain Fv region fused to *Escherichia coli* alkaline phosphatase, *Annals of the New York Academy of Science*, **646**, 106–114.

KÖHLER, G. and MILSTEIN, C., 1975, Continous cultures of fused cells secreting antibody of predefined specificity, *Nature*, **256**, 495–497.

KRETZSCHMAR, T., AOUSTIN, L., ZINGEL, O., MARANGI, M., VONACH, B., TOWBIN, H. and GEISER, M., 1996, High-level expression in insect cells and purification of secreted monomeric single-chain Fv antibodies, *Journal of Immunological Methods*, **195**, 93–101.

LEE, S.Y., 1996, High cell-density culture of *Escherichia coli*, *Trends in Biotechnology*, **14**, 98–105.

LILLEY, G.G., DOLEZAL, O., HILLYARD, C.J., BERNARD, C. and HUDSON, P.J., 1994, Recombinant single chain antibody peptide conjugates expressed in *Escherichia coli* for rapid diagnosis of HIV, *Journal of Immunological Methods*, **171**, 211–226.

LINDNER, P., GUTH, B., WÜLFING, C., KREBBER, C., STEIPE, B., MÜLLER, F. and PLÜCKTHUN, A., 1992, Purification of native proteins from cytoplasm and periplasm of *Escherichia coli* using IMAC and histidine tails: a comparison of proteins and protocols, *Methods: A Companion to Methods in Enzymology*, **4**, 41–56.

LINDNER, P., BAUER, K., KREBBER, A., NIEBA, L., KREMMER, E., KREBBER, C., *et al.*, 1997, Specific detection of his-tagged proteins with recombinant anti-His tag scFv-phosphatase or scFv-phage fusions, *Biotechniques*, **22**, 140–149.

LISSER, S. and MARGLAIT, H., 1993, Compilation of *E. coli* mRNA promoter sequences, *Nucleic Acids Research*, **21**, 1507–1516.

LUO, D., MAH, N., KRANTZ, M., WILDE, K., WISHART, D., ZHANG, Y., *et al.*, 1995, Vl-linker-Vh orientation-dependent expression of single chain Fv-containing an engineered disulphide-stabilized bond in the framework regions, *Journal of Biochemistry*, **118**, 825–831.

MAKRIDES, S.C., 1996, Strategies for achieving high-level expression of genes in *Escherichia coli*, *Microbiological Reviews*, **60**, 512–538.

MALINOWSKI, D.P., GOURLEY, M., EDELSTEIN, S. and PEARSON, R.E., 1992, Expression of a *Chlamydia* anticarbohydrate single-chain antibody as a maltose binding fusion protein, *Cell Biophysics*, **21**, 1–12.

MANDECKI, W., SHALLCROSS, M.A., SOWADSKI, J. and TOMAZIC-ALLEN, S., 1991, Mutagenesis of conserved residues within the active site of *Escherichia coli* alkaline phosphatase yields enzymes with increased k_{cat}, *Protein Engineering*, **4**, 801–804.

MARKS, J.D., HOOGENBOOM, H.R., BONNERT, T.P., MCCAFFERTY, J., GRIFFITHS, A.D. and WINTER, G., 1991, By-passing immunization: human antibodies from V-gene libraries displayed on phage, *Journal of Molecular Biology*, **222**, 581–597.

MCCAFFERTY, J., GRIFFITHS, A.D., WINTER, G. and CHISWELL, D.J., 1990, Phage antibodies: filamentous phage displaying antibody variable domains, *Nature*, **348**, 552–554.

MCCARTNEY, J.E., LEDERMAN, L., DRIER, E.A., CABRAL-DENISON, N.A., WU, G.-M., BATORSKY, R.S., *et al.*, 1991, Biosynthetic antibody binding sites: development of a single-chain Fv Model based on antidinitrophenol IgA Myeloma MOPC 315, *Journal of Protein Chemistry*, **10**, 669–683.

MCCLURE, W.R., 1985. Mechanism and control of transcription initiation in prokaryotes, *Annual Review of Biochemistry*, **54**, 171–204.

MCGREGOR, D.P., MOLLOY, P.E., CUNNINGHAM, C. and HARRIS, W.J., 1994, Spontaneous assembly of bivalent single chain antibody fragments in *Escherichia coli*, *Molecular Immunology*, **31**, 219–226.

NERI, D., DE LALLA, C., PETRUL, H., NERI, P. and WINTER, G., 1995, Calmodulin as a versatile tag for antibody fragments, *Biotechnology*, **13**, 373–377.

NIEBA, L., HONEGGER, A., KREBBER, C. and PLÜCKTHUN A., 1997, Disrupting the hydrophobic patches at the antibody variable/constant domain interface: improved *in vivo* folding and physical characterization of an engineered scFv fragment, *Protein Engineering*, **10**, 435–444.

NILSON, B.H.K., LÖGDBERG, L., KASTERN, W., BJÖRK, L. and ÅKERSTRÖM, B., 1993, Purification of antibodies using protein L-binding framework structures in the light chain variable domain, *Journal of Immunological Methods*, **164**, 33–40.

NYYSSÖNEN, E., PENTTILAE, M., HARKKI, A., SALOHEIMO, A., KNOWLES, J.K.C. and KERAENEN, S., 1993, Efficient production of antibody fragments by the filamentous fungus *Trichoderma reesei*, *Biotechnology*, **11**, 591–595.

OPPENHEIM, A.B., GILADI, H., GOLDENBERG, D., KOBI, S. and AZAR, I., 1996, Vectors and transformed host cells for recombinant protein production at reduced temperatures, International patent application WO 96/03521.

PACK, P. and PLÜCKTHUN, A., 1992, Miniantibodies: use of amphipathic helices to produce functional, flexibly linked dimeric Fv Fragments with high avidity in *Escherichia coli*, *Biochemistry*, **31**, 1579–1584.

PACK, P., KUJAU, M., SCHROECKH, V., KNÜPFER, U., WENDEROTH, R., RIESENBERG, D. and PLÜCKTHUN, A., 1993, Improved bivalent miniantibodies, with identical avidity as whole antibodies, produced by high cell density fermentation of *Escherichia coli*, *Biotechnology*, **11**, 1271–1277.

PANTOLIANO, M.W., BIRD, R.E., JOHNSON, S., ASEL, A.D., DODD, S.W., WOOD, J.F. and HARDMAN, K.D., 1991, Conformational stability, folding, and ligand-binding affinity of single-chain Fv immunoglobulin fragments expressed in *Escherichia coli*, *Biochemistry*, **30**, 10117–10125.

PLATT, T., 1986, Transcriptional termination and the regulation of gene expression, *Annual Review of Biochemistry*, **55**, 339–372.

PLÜCKTHUN, A. and PACK, P., 1997, New protein engineering approaches to multivalent and bispecific antibody fragments, *Immunotechnology*, **3**, 83–105.

PLÜCKTHUN, A. and SKERRA, A., 1989, Expression of functional antibody Fv and Fab fragments in *Escherichia coli*, *Methods in Enzymology*, **178**, 497–515.

POWER, B.E., IVANCIC, N., HARLEY, V.R., WEBSTER, R.G., KORTT, A.A., IRVING, R.A. and HUDSON, P.J., 1992, High-level temperature-induced synthesis of an antibody V_H-domain in *Escherichia coli* using the PelB secretion signal, *Gene*, **113**, 95–99.

PROBA, K., GE, L.M. and PLÜCKTHUN, A., 1995, Functional antibody single chain fragments from the cytoplasm of *Escherichia coli*: influence of thioredoxin reductase (TrxB), *Gene*, **159**, 203–207.

PUGSLEY, A.P., 1993, The complete general secretory pathway in Gram-negative bacteria, *Microbiological Reviews*, **57**, 50–108.

RICHARDSON, J.P. and GREENBLATT, J., 1996, Control of RNA chain elongation and termination, in NEIDHARDT F.C., CURTISS III, R., INGRAHAM, J.L., *et al.*, (Eds), *Escherichia and Salmonella: Cellular and Molecular Biology*, Vol. 1, pp. 822–848, Washington: ASM Press.

RIDDER, R., SCHMITZ, R., LEGAY, F. and GRAM, H., 1995, Generation of rabbit monoclonal antibody fragments from a combinatorial phage display library and their production in the yeast *Pichia pastoris*, *Biotechnology*, **13**, 255–260.

RINGQUIST, S., SHINEDLING, S., BARRICK, D., GREEN, L., BINKLEY, J., STORMO, G.D. and GOLD, L., 1992, Translation initiation in *Escherichia coli*: sequences within the ribosome-binding site, *Molecular Microbiology*, **6**, 1219–1229.

SAMBROOK, J., FRITSCH, E.F. and MANIATIS, T., 1989, *Molecular Cloning, A Laboratory Manual*, 2nd edition, Cold Spring Harbor: Cold Spring Harbor Laboratory Press.

SAWYER, J.R., TUCKER, P.W. and BLATTNER, F.R., 1992, Metal-binding chimeric antibodies expressed in *Escherichia coli*, *Proceedings of the National Academy of Sciences of the USA*, **89**, 9754–9758.

SCHEIN, C.H., 1993, Solubility and secretability, *Current Opinion in Biotechnology*, **4**, 456–461.

SCHIER, R., BYE, J., APELL, G., MCCALL, A., ADAMS, G.P., MALMQUIST, M., *et al.*, 1996, Isolation of high affinity monomeric human anti-c-*erb* B2 single chain Fv using affinity-driven selection, *Journal of Molecular Biology*, **255**, 28–43.

SCHMIDT, T.G.M. and SKERRA, A., 1993, The random peptide library-assisted engineering of a C-terminal affinity peptide useful for the detection and purification of a functional Ig Fv fragment, *Protein Engineering*, **6**, 109–122.

SCHMIDT, T.G. and SKERRA, A., 1994, One-step affinity purification of bacterially produced proteins by means of the Strep tag and immobilized recombinant core streptavidin, *Journal of Chromatography A*, **676**, 337–345.

SCHOUTEN, A., ROOSIEN, J., DE BOOR, J.M., WILMINK, A., ROSSO, M.N., BOSCH, D., *et al.*, 1997, Improving scFv antibody expression levels in the plant cytosol, *FEBS Letters*, **415**, 235–241.

SHEETS, M.D., AMERSDORFER, P., FINNERN, R., SARGENT, P., LINDQVIST, E., SCHIER, R., *et al.*, 1998, Efficient construction of a large non-immune phage antibody library: The production of high-affinity human single-chain antibodies to protein antigens, *Proceedings of the National Academy of Sciences of the USA*, **95**, 6157–6162.

SHINE, J. and DALGARNO, L., 1974, The 3′-terminal sequence of *Escherichia coli* 16S ribosomal RNA: complementarity to nonsense triplets and ribosome binding sites, *Proceedings of the National Academy of Sciences of the USA*, **71**, 1342–1346.

SIMMONS, L.C. and YANSURA, D.G., 1996, Translational level is a critical factor for the secretion of heterologous proteins, in *Escherichia coli*, *Nature Biotechnology*, **14**, 629–634.

SKERRA, A., 1993, Bacterial expression of immunoglobulin fragments, *Current Opinion in Immunology*, **5**, 256–362.

SKERRA, A., 1994a, A general vector, pASK84, for cloning, bacterial production, and single-step purification of antibody Fab fragments, *Gene*, **14**, 79–84.

SKERRA, A., 1994b, Use of the tetracycline promoter for the tightly regulated production of a murine antibody fragment in *Escherichia coli*, *Gene*, **151**, 131–135.

Skerra, A. and Plückthun, A., 1988, Assembly of a functional immunoglobulin Fv fragment in *Escherichia coli*, *Science*, **240**, 1038–1041.

Skerra, A. and Plückthun, A., 1991, Secretion and *in vivo* folding of the Fab fragment of the antibody McPC603 in *Escherichia coli*: influence of disulphides and cis-prolines, *Protein Engineering*, **4**, 971–979.

Smith, G.P., 1988, Filamentous phages as cloning vectors, in Rodriguez, R.L. and Denhardt, D.T. (Eds) *Vectors: A Survey of Molecular Cloning Vectors and Their Uses*, pp. 61–83, Boston: Betterworths.

Takagi, H., Morinaga, Y., Tsuchiya, M., Ikemura, H. and Inouye, M., 1988, Control of folding of proteins secreted by a high expression secretion vector, pIN-III-*omp*A: 16-fold increase in production of active subtilisin E in *Escherichia coli*, *Biotechnology*, **6**, 948–950.

Thomann, H.-U., Ibba, M., Hong, K.-W. and Söll, D., 1996, Homologous expression and purification of mutants of an essential protein by reverse epitope-tagging, *Biotechnology*, **14**, 50–55.

Torrance, L., 1992, Serological methods to detect plant viruses: production and use of monoclonal antibodies, in Duncan, J.M. and Torrance, L. (Eds) *Techniques for the Rapid Detection of Plant Pathogens*, pp. 7–33, Oxford: Blackwell.

Turner, D.J., Ritter, M.A. and George, A.J., 1997, Importance of the linker in expression of single chain Fv antibody fragments: optimisation of peptide sequence using phage display technology, *Journal of Immunological Methods*, **205**, 43–54.

Van Regenmortel, M.H.V., 1982, *Serology and Immunochemistry of Plant Viruses*, New York: Academic Press.

Vieira, J. and Messing, J., 1987, Production of single-stranded plasmid DNA, *Methods in Enzymology*, **153**, 3–11.

von Strandmann, E.P., Zoidl, C., Nakhei, H., Holewa, B., von Strandmann, R.P., Lorenz, P., *et al.*, 1995, A highly specific and sensitive monoclonal antibody detecting histidine-tagged recombinant proteins, *Protein Engineering*, **8**, 733–735.

Voorma, H.O., 1996, Control of translation initiation in prokaryotes, in Hershey, J.W.B., Mathews, M.B. and Sonenberg, N. (Eds) *Translational Control*, pp. 759–777, Cold Spring Harbor, NY: Cold Spring Harbor Laboratory Press.

Wall, J.G. and Plückthun, A., 1995, Effects of overexpressing folding modulators on the *in vivo* folding of heterologous proteins in *Escherichia coli, Current Opinion in Biotechnology*, **6**, 507–616.

Wang, L.F., Yu, M., White, J.R. and Eaton, B.T., 1996, Btag: a novel six-residue epitope tag for surveillance and purification of recombinant proteins, *Gene*, **169**, 53–58.

Ward, E.S., Güssow, D., Griffiths, A.D., Jones, P.T. and Winter, G., 1989, Binding activities of a repertoire of single immunoglobulin variable domains secreted from *Escherichia coli*, *Nature*, **341**, 544–546.

Ward, V.K., Kreissig, S.B., Hammock, B.D. and Choudary, P.V., 1995, Generation of an expression library in the baculovirus expression vector system, *Journal of Virological Methods*, **53**, 263–272.

Weiss, E. and Van Regenmortel, M.H., 1989, Use of rabbit Fab′-peroxidase conjugates prepared by the maleimide method for detecting plant viruses by ELISA, *Journal of Virological Methods*, **24**, 11–25.

Weiss, E. and Orfanoudakis, G., 1994, Application of an alkaline phosphatase fusion protein system suitable for efficient screening and production of Fab-enzyme conjugates in *E. coli*, *Journal of Biotechnology*, **33**, 43–53.

Weiss, E., Chatellier, J. and Orfanoudakis, G., 1994, *In vivo* biotinylated recombinant antibodies: construction, characterization, and application of a bifunctional Fab-BCCP fusion protein produced in *Escherichia coli*, *Protein Expression and Purification*, **5**, 509–517.

Wels, W., Harwerth, I.-M., Zwickl, M., Hardman, N., Groner, B. and Hynes, N.E., 1992, Construction, bacterial expression and characterization of a bifunctional single-chain antibody-phosphatase fusion protein targeted to the human erbB-2 receptor, *Biotechnology*, **10**, 1128–1132.

WU, X.C., NG, S.C., NEAR, R.I. and WONG, S.L., 1993, Efficient production of a functional single-chain antidigoxin antibody via an engineered *Bacillus subtilis* expression-secretion system, *Biotechnology (NY)*, **11**, 71–76.

ZIEGLER, A., TORRANCE, L., MACINTOSH, S.M., COWAN, G.H. and MAYO, M., 1995, Cucumber mosaic cucumovirus antibodies from a synthetic phage display library, *Virology*, **214**, 235–238.

5

Analysis of Antibody Binding Properties

GABRIELE ZEDER-LUTZ AND MARC H.V. VAN REGENMORTEL

5.1 Introduction

A wide repertoire of antibody specificities can be generated by the immune system or with the use of recombinant technology. These antibodies find many applications in medicine and are also useful tools for fundamental studies in immunology. The usefulness of each antibody depends on its antigenic specificity as well as on its binding properties. Antibody–antigen interaction can be analysed at different levels of resolution going from the footprint of the antibody molecule on the antigen surface to the atomic scale at the level of amino acid residues involved in the binding. Affinity and kinetic measurements are able to reveal the energetic contribution of the residues of each partner to the binding.

Once epitopes have been localized at the antigen surface by a procedure known as epitope mapping, it is easier to select antibody probes suitable for diagnostic assays. The ELISA technique using peptides adsorbed on microtitre plates is currently widely used for the diagnosis of virus infections (Leinikki *et al.*, 1993). Epitope mapping is also useful for analysing any biological function such as enzyme or hormone activity which is linked to the region of the protein where epitopes are located (Alonso-Whipple *et al.*, 1988; Daiss and Scalice, 1994; Daiss *et al.*, 1995). In the case of viruses, for instance, the identification of neutralization epitopes may be of considerable help for developing efficient synthetic vaccines.

The relative position of epitopes at the surface of an antigen can be established with a panel of antibodies using two-site immunoassays (Daiss and Scalice, 1994; Saunal and Van Regenmortel, 1995a). A more precise localization of epitopes can be achieved by competitive assays using natural (Smith-Gill *et al.*, 1984; Harper *et al.*, 1987) or recombinant antigen homologues and chemically modified analogues (Quesniaux *et al.*, 1987; Alexander *et al.*, 1992; Jin and Wells, 1994), as well as by peptide scanning (Geysen *et al.*, 1984; Lane and Stephen, 1993; Wang *et al.*, 1995) and various types of structural analysis (Paterson, 1992; Padlan, 1994).

Additional insight in to the nature of antigen–antibody interactions can be obtained by analysing the binding in terms of affinity and kinetic constants which provide information about the strength and speed of the reaction. The affinity of antibodies determines to a considerable degree their behaviour in immunoassays and affinity ranking of monoclonal antibodies is helpful for predicting their potential usefulness in a particular type of assay. Antibodies of higher affinity are able to detect smaller amounts of antigen (Steward and Lew, 1985). On the other hand, it is preferable to use antibodies of moderate affinity for the purification of antigens by affinity chromatography since such antibodies will make it possible to elute the antigen under mild conditions. Any observed change in the affinity of an antibody points to alterations in the structure of one or both reactants. Measurements of the affinity of a panel of monoclonal antibodies (Mabs) are thus useful for characterizing recombinant proteins and monitoring undesirable denaturation during protein purification schemes (Van Regenmortel, 1994; Mani *et al.*, 1997).

Antigen–antibody interactions may be described by a simple one-to-one interaction model:

$$A + B \underset{k_d}{\overset{k_a}{\rightleftharpoons}} AB$$

where A represents free antigen, B free antibody and AB the antigen–antibody complex. k_a and k_d are the association and dissociation rate constants respectively. At equilibrium, the rates of association and dissociation are equal and the equilibrium association constant K may be calculated from:

$$K = \frac{k_a}{k_d} = \frac{[AB]}{[A]\,[B]} = \frac{1}{K_d} \tag{5.1}$$

This constant K ($l\ mol^{-1}$) corresponds to the inverse of K_d ($mol\ l^{-1}$) which is equivalent to the free antibody concentration when 50 per cent of antigen is bound, i.e. when [A] = [AB]. In order to have the bound antigen varying from 10 to 90 per cent it is necessary to use total antibody concentrations that range from $0.1K_d$ to $10K_d$. Since the complexes form and dissociate continuously, the equilibrium is a dynamic state. Several methods can be used to quantify the antigen–antibody interaction at equilibrium (Azimzadeh and Van Regenmortel, 1991; Phizicky and Fields, 1995; Neri *et al.*, 1996) but there are only very few methods for kinetic analysis. At present, the easiest and most reliable way to measure binding rates is by means of biosensor instruments based on surface plasmon resonance.

This chapter will focus mainly on the analysis of antibody–antigen interactions using biosensor technology. General information on the mechanism of immunochemical interactions and on the various methods used in their analysis is available in several reviews (Van Oss and Van Regenmortel, 1994; Van Regenmortel, 1997a).

5.2 Principle and description of the BIAcore

The BIAcore technology monitors biomolecular interactions in real time without labelling any reactants (Malmqvist, 1993; Szasbo *et al.*, 1995). To perform an analysis, one interactant (the ligand) is immobilized on a sensor surface (the sensor chip) and the sample containing the other interactant (the analyte) is injected over the surface in a controlled flow of buffer by an automated fluid-handling system. The interaction between analyte and the immobilized ligand is monitored by the optical phenomenon of surface plasmon resonance. The instrument detects changes in refractive index, due to the accumulation of bound molecules onto the sensor surface. The sensor chip is a glass

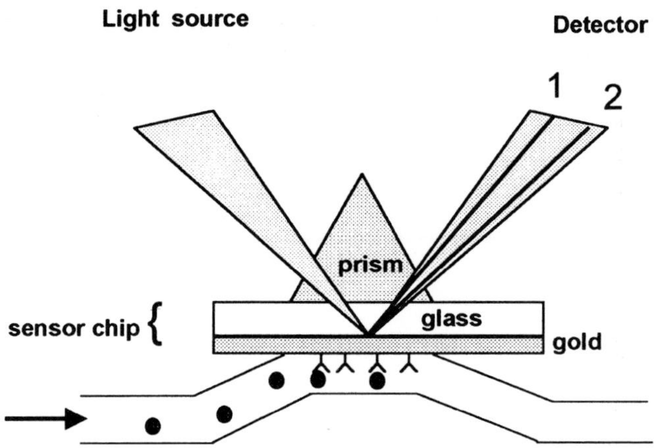

Figure 5.1 Real time surface plasmon resonance (SPR) detection of biomolecular interactions. The figure shows the configuration of the SPR detection system as well as with the sensor chip and the flow channel. When SPR occurs, the intensity of reflected light is minimized at a certain angle. This angle is deviated as a function of the refractive index in the vicinity of the surface and increases when analyte binds to the immobilized ligand.

slide covered with a thin gold film where a carboxylated dextran matrix has been covalently attached. The dextran matrix provides a hydrophilic environment for studying interactions, which prevents most of the unspecific binding. The matrix carboxyl groups provide a flexible anchor for immobilizing the ligand by a variety of chemistries (Johnsson *et al.*, 1995), and this allows interactions to occur as in solution (Johnsson *et al.*, 1991). The chip is held in contact with the prism of the optical system by the microfluidic cartridge (Figure 5.1). Monochromatic polarized incident light is focused by the prism and illuminates the glass side of the sensor chip. At a specific incident angle, the light travelling through a more optically dense medium is totally reflected into the prism when reaching a medium of lesser optical density. Although light is totally reflected, a component of the incident light, called the evanescent wave, penetrates into the buffer. If the interface between the two media is coated with a gold film, photons of the light will resonate with free oscillating electrons called plasmons. This phenomenon is called surface plasmon resonance (SPR). When SPR occurs, light energy is lost to the metal film and the reflected light intensity decreases at a sharply defined angle. This SPR angle is very sensitive to refractive index changes in the vicinity of the surface. Changes in refractive index, due for instance to the binding of analyte to the immobilized ligand, deviate the SPR angle position. By continuous monitoring of the SPR angle deviation, interactions can be followed in real time. The specific response is expressed in arbitrary resonance units (RU) which is almost the same for all proteins and peptides as well as for glycoproteins, lipids and nucleic acids; 1000 RU correspond to a 0.1° shift in the SPR angle and to a surface concentration change of about 1 ng/mm^2 (Stenberg *et al.*, 1991).

The continuous display of RU as a function of time (the corresponding plot is called a sensorgram) provides a complete record of the progress of the association and dissociation between two molecules (Figure 5.2). After analysis, non-covalently bound interactants may be removed from the sensor chip by appropriate regeneration so that the surface

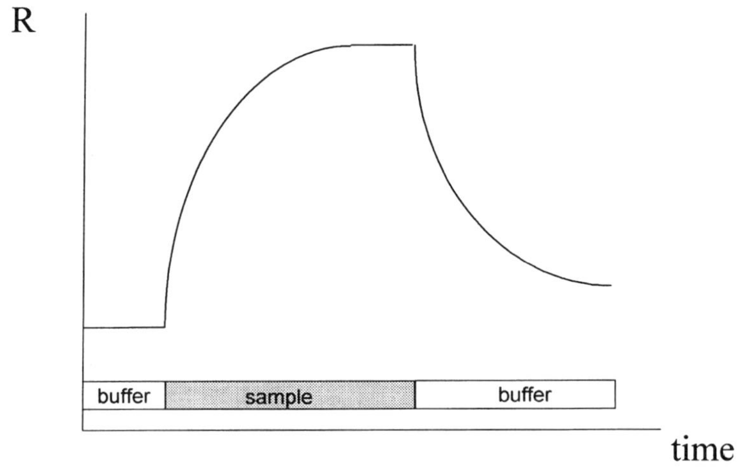

Figure 5.2 Sensorgram describing the response (R) recorded as a function of time. R is proportional to the mass of bound molecules. At the origin, the baseline corresponds to a flow of buffer over the immobilized ligand matrix. When the analyte is injected, the signal increases as molecules bind to the ligand. Dissociation of the complexes occurs when a buffer replaces the sample in the microfluidic tubing.

can be used for a new analysis. Depending on the stability of the immobilized ligand, the surface can be regenerated from 30 to more than 100 times.

5.3 Use of the BIAcore for epitope mapping

Epitope mapping using monoclonal antibodies is a powerful method for examining the surface topology of antigens and is useful for assessing their biological activity.

Various methods can be used for mapping epitopes (Morris, 1996; Van Regenmortel, 1996) but the most common technique consists of the pair-wise analysis of the binding of Mabs by radioimmunoassay (RIA) or enzyme-linked immunosorbent assay (ELISA). In such assays, one Mab is adsorbed to a solid phase, the antigen is bound and the ability of the second antibody to bind to the complex is tested. By repeated experiments all combinations of antibody pairs can be analysed. Pairs of Mabs that bind independently of each other are useful for sandwich immunoassays in which the first antibody captures the antigen and the second antibody, which is labelled, is used as detector.

Epitope mapping using the BIAcore Epitope mapping using the BIAcore follows essentially the same scheme, but has several major advantages. Since each step is visualized in real time, the behaviour of each successive reactant can be controlled, and no labelling of any reagent is necessary. Furthermore, since the detected response is proportional to the protein mass, each Mab–antigen binding event can be quantified and the stoichiometry of the complex can be calculated as the molar ratio (MR) of the reactant by the following relationship:

$$MR = (RU_{Mab}/RU_{Ag}) \cdot (MW_{Ag}/MW_{Mab}).$$

Expressing the data as MR has the advantage that SPR signals can be normalized. A threshold in response level or in MR has to be chosen as the criterion for differentiating binding from no binding (Daiss and Scalice, 1994). A high binding ratio (ideally MR = 1) of the second Mab to the capture antigen–antibody complex shows that the particular pair binds independently. However, a very low binding (MR about 0) of the second antibody can either result from a partially modified antigen or from an interfering reaction.

In practice, it is not exceptional to obtain partial MR ($0 \ll MR \ll 1$) for non-interfering interactions (Daiss *et al.*, 1995; Saunal and Van Regenmortel, 1995b). Several possible reasons may explain this phenomenon. One possible explanation is that the second Mab does not saturate the antigen because of poor experimental design (Daiss and Scalice, 1994). It is important to consider that during binding assays continuous dissociation and reassociation of the binding partners occur so that the response is also dependent on the kinetic properties of the interaction. It is clear from theoretical simulations that analyte concentration as well as the association rate will determine the time necessary to reach antigen saturation (Figure 5.3A). A rapid dissociation ($k_d > 10^{-3}\ s^{-1}$) may cause dissociation of the complex between the first Mab and antigen before the second Mab has been injected (Figure 5.3B). Visual inspection of the sensorgram may indicate whether this is due to too low concentration and insufficient injection time of the second Mab, or to the use of a low affinity Mab which is unable to saturate the antigen. In order to avoid too low coverage of antigen it is important to use a first Mab with stable binding properties ($k_d < 10^{-3}\ s^{-1}$) and a concentration that is at least ten times higher than the K_d of the second Mab (Malmqvist, 1996). Other possible factors may explain low stoichiometry of binding such as bivalent binding of Mab, steric hindrance within the carboxylated matrix (Daiss and Scalice, 1994) or steric hindrance resulting from two Mabs binding to epitopes located very close to each other (Saunal and Van Regenmortel, 1995b). In pair-wise assays, changing the order of the reagents to form the complex can give different results for the same pair of Mabs (so-called asymmetric interactions). Therefore, in case of a negative result, the pair of Mabs should always be tested also in the reverse order (Fägerstam *et al.*, 1990). Different interpretations have been suggested for non-reciprocal interactions such as the magnitude of the kinetic constants, heterogeneity of antigen molecules (Daiss and Scalice, 1994) and modification of the conformational state of the antigen upon binding to the first Mab, which leads to the disappearance of the epitope recognized by the second Mab.

For antigens with multiple subunits, the interpretation can be more complicated, since saturation of all first Mab sites on the antigen may be difficult to achieve. This could then lead to secondary binding and to potentially false positives by non-interfering Mab pairs (Daiss and Scalice, 1994).

Pair-wise mapping can be performed either by immobilizing the first Mab directly to the sensor chip or by capturing it with rabbit anti-mouse Fc or subclass-specific antibodies (Fägerstam *et al.*, 1990). This second surface functionalization has the advantage of providing a stable Mab binding which is reversible by elution with 10 to 100 mM HCl. Furthermore, the first Mab, antigen and second antibody can be introduced directly from culture supernatant or other complex mixtures. The BIAcore technique also allows multi-site specificity tests using a sequence of several Mabs to confirm pair-wise mapping results (Fägerstam *et al.*, 1990; Daiss and Scalice, 1994). The first report of epitope mapping with BIAcore described a study of HIV-1 core protein p24 using a panel of 30 Mabs (Fägerstam *et al.*, 1990). A total of 17 groups of Mab showing the same reactivity pattern could be distinguished. In this study, four Mabs were unable to bind antigen when used as first antibody. Examination of sensorgrams showed that these Mabs lost their antigen binding capacity when captured by the first anti-immunoglobulin reagent immobilized on the chip, although they were active when used as second antibody. Without this information, the absence of signal for the second antibody in a classical two-site binding assay

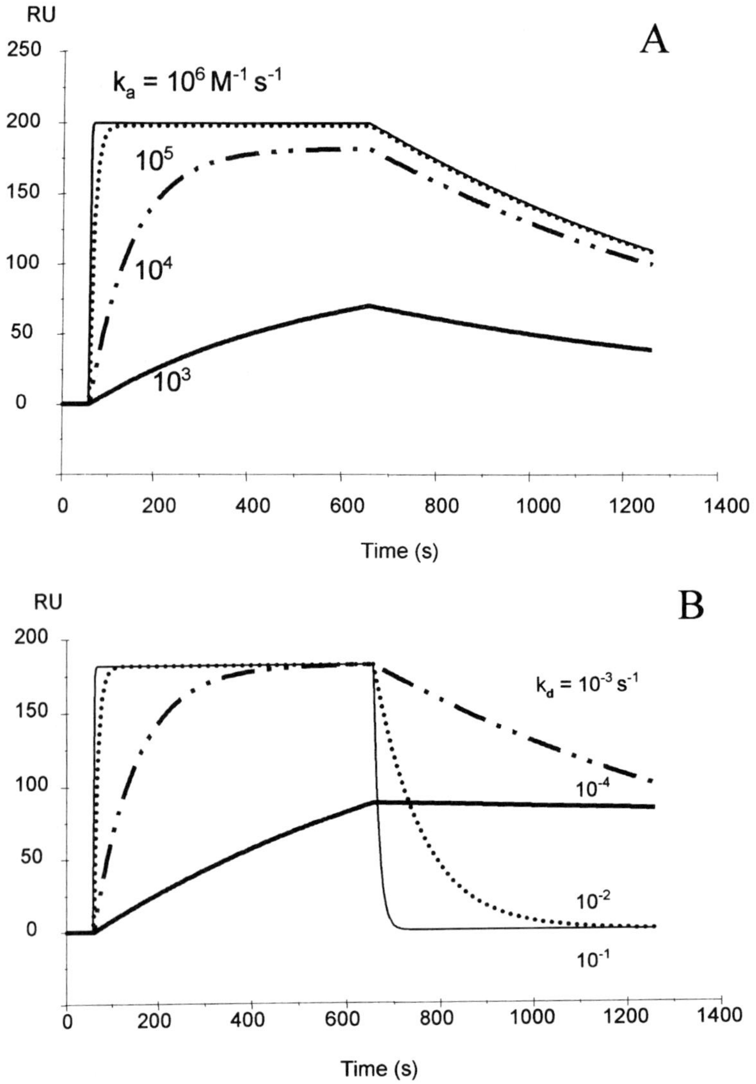

Figure 5.3 Simulated sensorgrams for the interaction of four antibodies at a concentration of 1 μM on an antigen surface of maximal binding capacity of 200 RU. (A) The Mabs have the same dissociation rate (10^{-3} s^{-1}) but different association rates. It can be seen that saturation of the antigen is reached quickly for the two high affinity antibodies while the antibody with the lowest k_a does not saturate the antigen during the 10 min of injection. (B) The Mabs have the same equilibrium association constant ($K = 10^7$ M^{-1}) but different dissociation rates. This example shows that a high dissociation rate ($k_d > 10^{-3}$ s^{-1}) of the complex will cause dissociation of the first Mab–antigen complex before the second antibody has been injected in a double antibody sandwich assay.

would have been misinterpreted as due to hindrance in the binding of the second Mab, rather than absence of bound antigen.

In another study, the epitopes of recombinant granulocyte colony stimulating factor were analysed with a panel of 15 Mabs (Nice *et al.*, 1993). Three epitope clusters were detected which agreed with earlier competition ELISA studies with biotinylated antibodies and peptide fragments.

The epitopes of *Tobacco mosaic virus* (TMV) (genus *Tobamovirus*) and its coat protein (TMVP) have also been analysed extensively by biosensor technology (Van Regenmortel *et al.*, 1993, 1997). Epitopes on viral coat proteins are difficult to analyse because of their tendency to form oligomers (Van Regenmortel, 1997). Polymerization of viral subunits into virions will result in some portions of the subunit surface becoming buried, and this leads to the disappearance of certain epitopes. Such epitopes have been called *cryptotopes* (Jerne, 1960). Furthermore, the polymerization of viral proteins brings some residues of neighbouring subunits into close juxtaposition, and these may be recognized as a single entity by certain antibodies. Such epitopes that arise through the quaternary structure of the protein have been called *neotopes* (Van Regenmortel, 1966). A third category of epitope present in both dissociated viral protein and intact virions has been called *metatope* (Van Regenmortel, 1966). It has been possible to locate some cryptotopes in the TMVP molecule because the corresponding anti-cryptotope antibodies reacted with natural or synthetic fragments of TMVP (Benjamini, 1977; Van Regenmortel, 1986; Trifilieff *et al.*, 1991; Van Regenmortel *et al.*, 1993). This kind of mapping was not feasible in the case of neotopes and metatopes because none of the corresponding antibodies reacted with linear fragments of TMVP. Immunoelectron microscopy gave some information on a limited number of neotopes and metatopes present on TMV particles (Dore *et al.*, 1988, 1990). All examined anti-metatope Mabs were found to react with the virion extremity containing the 5′ end of the RNA (surface E of the subunit), whereas the anti-neotope Mabs reacted along the entire length of the virus particle (surface A of the subunit). Whereas attempts to map anti-metatope antibodies by means of ELISA had been unsuccessful (Dekker *et al.*, 1989), initial experiments with BIAcore led to the identification of two non-overlapping metatopes in TMVP (Dubs *et al.*, 1992). Since BIAcore allows the stoichiometry of any interaction to be calculated, it was easy to determine whether antibodies bound along the entire surface of the TMV particle or only at one end of the particle representing 1.5 per cent of the total virion surface. Location of neotopes and metatopes was investigated by a double antibody capture assay (Saunal and Van Regenmortel, 1995a). The binding of anti-neotope and anti-metatope antibodies to TMV captured by a first Mab (anti-neotope 253P) is illustrated in Figure 5.4. When antibodies bind along the entire surface of the particles the molar ratio of bound Mab/TMV is about 300 (Figure 5.4A), whereas it is only about 10 when antibodies bind to only one extremity of the TMV rods (Figure 5.4B). In this study, the biosensor analysis made it possible to demonstrate the presence of neotope and metatope specificities on parts of the viral surface where they had not been found before. Two types of anti-metatope and anti-neotope Mabs reacting with different surfaces of the subunit could be distinguished. Some Mabs that react with the extremity of TMV containing the 5′ end of the RNA were able to block *in vitro* the cotranslational disassembly of TMV by ribosomes while other antibodies could not (Saunal *et al.*, 1993). As no correlation could be established between the binding kinetics or the affinity of the Mabs and their inhibitory capacity, it seemed likely that the functional activity shown by some of the antibodies was related to the precise location of their complementary epitopes. Mapping studies showed that Mabs possessing the strongest inhibitory capacity bound to the part of the surface E closest to the central axis of polymerized TMVP (Figure 5.5). As this part of the subunit is known to interact with the viral RNA, it was concluded that the Mabs prevent the interaction between virus and ribosomes by a steric hindrance phenomenon (Saunal and Van Regenmortel, 1995b).

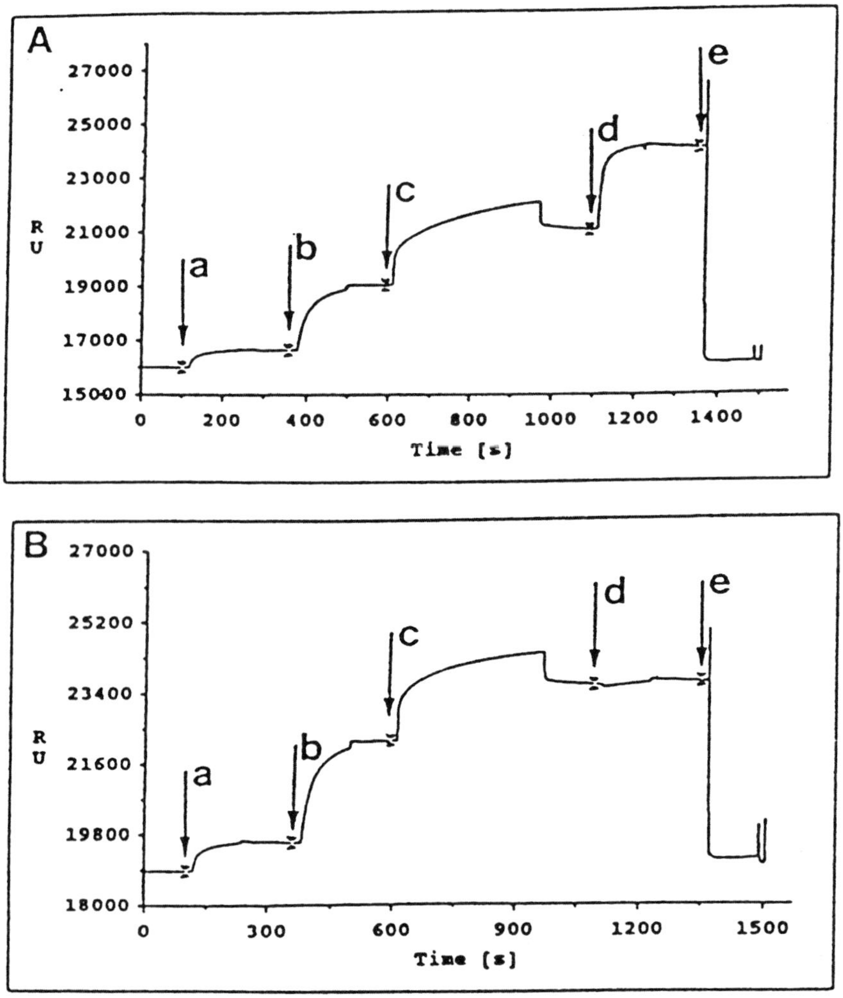

Figure 5.4 Sensorgram used for determining the binding sites of anti-neotope and anti-metatope antibodies on TMV particles. Phase (a) corresponds to the injection of capture anti-neotope Mab 253P. About 700 RU were bound. Phase (b) corresponds to TMV (100 μg/ml) injection (about 2500 RU are captured). Phase (c) corresponds to the saturation of rabbit anti-mouse globulins by means of non-specific ascitic fluid (about 1500–2500 RU bound). In A, phase (d) corresponds to the binding of Mab 5V. About 300 antibody molecules are bound per particle. In B, phase (d) corresponds to the binding of Mab 25P to only one extremity of the rods. About ten antibody molecules are bound per particle. Phase (e): the sensor surface was regenerated with a pulse of 100 mM HCl (from Saunal and Van Regenmortel, 1995b).

5.4 Measurement of affinities and kinetic constants of antibodies

Knowledge of antibody affinity is important for selecting suitable antibodies to be used in immunoassays. Moreover, the identification of residues that make up the combining sites makes it possible to modify the affinity and specificity of these sites by genetic engineering (Borrebaeck, 1995).

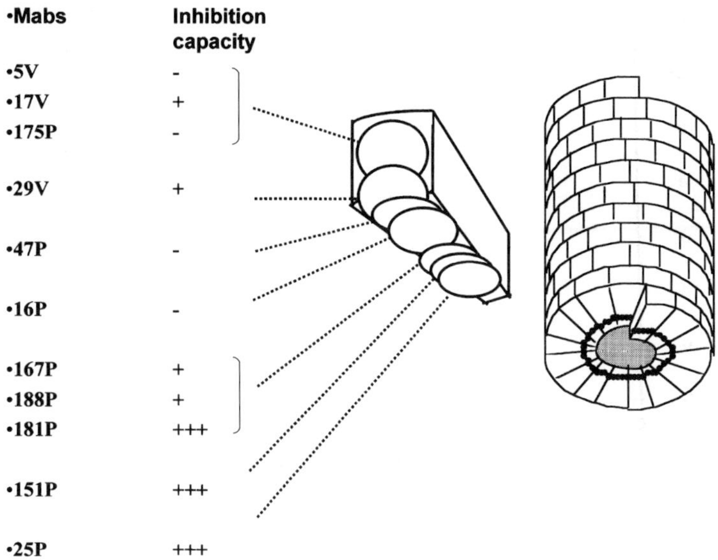

Figure 5.5 Schematic epitope map of TMVP constructed from pair-wise Mab assays using the BIAcore. The three antibodies 181P, 151P and 25P which strongly inhibit the cotranslational disassembly of TMV *in vitro* bind to the surface of the subunit that is in contact with the RNA (Saunal and Van Regenmortel, 1995a).

Numerous techniques exist to estimate the binding constants of protein–protein interactions (Phizicky and Fields, 1995; Lakey and Raggett, 1998). The sensitivity of each assay will determine the range of affinities that can be resolved and, for kinetic measurements, the time–resolution of the method will also be crucial.

5.4.1 *Affinity constant measurements*

The various experimental approaches for measuring the affinity constant of antigen–antibody interactions (Van Regenmortel and Azimzadeh, 1994; Neri *et al.*, 1996; Saunal *et al.*, 1997) can be classified in two groups depending on whether the equilibrium is reached at the liquid–solid interphase (heterogeneous phase techniques) or in solution (homogeneous phase techniques).

Heterogeneous phase techniques

In these methods, one of the reactants, usually the antigen, is adsorbed on a solid phase and free antibody is separated by simple washing. Some of the experimental limitations of this approach are that the passive immobilization of one partner on the plastic phase may alter its conformational integrity, and that a high density of immobilized molecules on the support may prohibit free diffusion of the partner in solution due to rebinding effects (Butler, 1992; Goldberg and Djavadi-Ohaniance, 1993). When the affinity of antibodies measured by solid-phase ELISA assay was compared with the values obtained in a liquid–phase test, higher K values were found in the solid phase assay (Nygren *et al.*, 1986; Azimzadeh *et al.*, 1992). This is probably due to the lower kinetic rate constants present in solid phase tests (Nygren *et al.*, 1986).

Table 5.1 Approximate calculated times required to reach 99.9 per cent of the steady state level at analyte concentration ranging from 0.001 to 1000 × K_d

Analyte concentration	k_d (s^{-1})			
	10^{-1}	10^{-2}	10^{-3}	10^{-4}
$K_d \times 10^{-3}$	69 s	11.5 min	115 min	19 h
$K_d \times 10^{-2}$	68.4 s	11.4 min	114 min	19 h
$K_d \times 10^{-1}$	62.8 s	10.5 min	104 min	17 h
K_d	34.5 s	6 min	58 min	9.6 h
$K_d \times 10^{1}$	6.3 s	1 min	10.5 min	105 min
$K_d \times 10^{2}$	0.7 s	7 s	1.1 min	11.4 min
$K_d \times 10^{3}$	0.7 s	7 s	68.4 s	69 s

Quantitative steady state measurements are only feasible for interactions with k_d of 10^{-3} s^{-1} or faster since the analyte cannot be injected longer than 12 h.

Real time interaction analysis by BIAcore also requires that one of the binding partners is immobilized. However, the hydrophilic and flexible nature of the dextran polymer present on the sensor chip matrix allows the analyte to bind very much as if it were in solution (Johnsson *et al.*, 1991). The equilibrium constant can be determined by analysing the concentration dependence of the equilibrium plateau response (Ladbury *et al.*, 1995; Dall'Acqua *et al.*, 1996). This approach has the advantage that it does not require modelling of the binding process (Ward *et al.*, 1995), and that it can be applied to interactions that have very fast binding kinetics (Ladbury *et al.*, 1995). However, the injection time required to reach equilibrium is a limiting factor. At a flow of 1 µl/min the maximum injection time that can be used is about 12 h in the BIAcore 2000 instrument. Time to reach equilibrium at a given concentration is only dependent on the k_d value (Table 5.1). Quantitative steady state measurements are thus only feasible for interactions with k_d values around 10^{-3} s^{-1} or faster. Alternative experimental protocols such as competitive binding have been proposed for equilibrium affinity measurements (Karlsson *et al.*, 1994; Nieba *et al.*, 1996).

Homogeneous phase techniques

In these methods, equilibrium between the bound and free molecules is reached in solution. Isothermal titration calorimetry (ITC) and spectrofluorometry allow direct measurements of the affinity while other methods require the physical separation of the free and bound molecules in order to measure the concentration of one of the reactants. The ELISA titration assay is an indirect separation technique (Friguet *et al.*, 1985) since the equilibrium between antigen and antibody molecules is pre-established in solution and free antibody concentration at equilibrium is titrated by ELISA.

Isothermal titration calorimetry (ITC) This method measures the heat produced or consumed during the binding reaction, which is proportional to the amount of complex formed (Doyle, 1997). One reactant (usually the antibody) is titrated and the heat produced or consumed in each titration step is quantified. The major advantage of ITC is that it gives information on the equilibrium association constant K, the enthalpy change ΔH, and the stoichiometry of a binding reaction in a single experiment without chemically altering the reaction partners. Using the new generation of titration calorimeters, heat changes as small as 10^{-7} cal s^{-1} can be accurately measured using nanomolar quantities of reactants. One limitation of this approach is that affinity values higher than 10^{8} M^{-1} cannot be determined (Wiseman *et al.*, 1989).

Spectrofluorometric methods In one kind of assay, an altered fluorescence spectrum for one of the interactants is measured upon formation of the antigen–antibody complex. In some cases, the natural fluorescence of tryptophan residues in the antibody can be used, but it is also possible to label either the antigen or the antibody with a fluorophore. Changes in the fluorescence emission spectrum upon complex formation can lead either to a shift in the wavelength of maximum fluorescence emission or to a change in fluorescence intensity (quenching or enhancement of fluorescence). It is then possible to evaluate the equilibrium constant K by measuring the fluorescence at a particular wavelength. If fluorescence changes are followed over time, kinetic parameters can also be determined. A stopped flow instrument for rapid mixing of the reactants is then required.

Fluorescence polarization is another technique based on fluorescent probes that can be used to determine binding affinity. This technique monitors the rotational motion of molecules and provides polarization values which increase as complexes are formed. The affinity is determined by titrating the labelled antigen with a series of antibody concentrations (Checovich *et al.*, 1995).

A major advantage of these methods is their high sensitivity which renders them highly suitable for measuring high affinity interactions ($K > 10^{9}$ M^{-1}). Both reactants are free to diffuse as the interaction is followed in solution so that no problem of mass transfer occurs. However, one limitation of the technique is that it is usually necessary to label one of the two reactants. This drawback may be overcome by a recently developed technique which allows the biosynthetic incorporation of unnatural fluorescent amino acids at defined sites of proteins using codon suppression. Using this approach, Turcatti *et al.* (1996) were able to follow the binding of a peptide ligand to tachykinin neurokinin-2 receptor expressed on *Xenopus* oocyte membranes by fluorescence resonance energy transfer.

Separation methods These methods are effective only if the dissociation of the complex is slow compared with the time necessary to complete the separation process. The molecular weight of the antigen will determine the most adequate separation method. The technique of equilibrium dialysis is suitable for small hapten antigens since bound and unbound haptens are separated while equilibrium is being reached. Techniques such as electrophoresis, gel filtration, selective precipitation with ammonium sulphate, or chromatography can be used for both haptens and protein antigens. In these cases, the antigen is usually labelled with a radioactive or fluorescent probe. Ultracentrifugation can be used if there is a significant difference between the sedimentation coefficients of the antigen and the antibody (McInerney *et al.*, 1993). For large antigens such as viruses, centrifugation can be used to separate free from bound antibody. The free antibody concentration present in the supernatant after centrifugation can be determined, for instance, by an ELISA titration method using a calibration curve (Azimzadeh and Van Regenmortel, 1991).

ELISA titration This technique developed by Friguet *et al.* (1985) is a homogeneous phase assay since equilibrium is reached in solution, and the free antibody concentration is determined using a microtitre plate coated with antigen. Conditions have to be chosen to ensure that equilibrium in solution is not affected during the subsequent ELISA. The advantages of this sensitive method are that labelling procedures are avoided and that crude antibody sample can be used. The upper limit of the affinity values that can be determined by ELISA using chromogenic substrates is about 10^9–10^{10} M^{-1}. Fluorogenic substrates may further increase the sensitivity of the assay by a factor of 100 (Djavadi-Ohaniance *et al.*, 1986).

The method described above for measuring free antibody concentration has been adapted for use with BIAcore (Zeder-Lutz *et al.*, 1993). A constant amount of antigen is mixed with increasing amounts of antibody and incubated to reach equilibrium. The mixture is injected over a sensor chip with a high density of immobilized antigen. Free antibody concentration is determined for each mixture using a calibration curve established with serial dilutions of known antibody concentration. The sensitivity of the biosensor makes it possible to determine even low antibody concentrations (0.1 nM) and a large range of affinities (K_d ranging from 10 mM to 1 nM).

Analysis of the binding data The binding data are analysed by means of Scatchard plots. Equation 5.1 can be expressed in terms of total antibody and antigen concentration used in the reaction. The association equilibrium K of an antibody may be expressed in terms of the mass action law:

$$K = \frac{sx}{(As - sx)(Bn - sx)} = \frac{ny}{(As - ny)(Bn - ny)} \tag{5.2}$$

where A = total antigen concentration (mol l^{-1}); B = total antibody concentration (mol l^{-1}); s = antigen valence; n = antibody valence; y = bound antibody concentration (mol l^{-1}); Bn = total antibody sites (mol l^{-1}); and sx = ny = total of bound sites. The concentration of bound sites can be calculated by solving a quadratic equation. From the values of A and B, the percentage of molecules engaged in the complex can be calculated and equation 5.2 can be used to derive K values from the percentage of antibody complexed at given values of A and B, using computer fitting software. Alternatively, equation 5.2 can be linearized in a suitable form for Scatchard representation of binding data:

$$f/d = K\,(s - nf)$$

where f is the ratio of bound antibody to total antigen, d is the free antibody concentration and s and n are the antigen and antibody valencies respectively. The Scatchard plot of f/d versus nf gives the value of K as the slope and the value of the antigenic valence by extrapolation of the x-axis at the origin (Van Regenmortel, 1997b).

5.4.2 *Kinetic constants measurements*

While the quantities of complex formed depend on the equilibrium affinity constant K, the time necessary to reach equilibrium depends on the kinetic rates. In fact, the physiological relevance of many biological processes lies in their kinetic rates rather than their equilibrium affinities. It has been shown, for instance, that the dissociation rate of anti-gp120 antibodies is predictive of their virus neutralization capacity (Van Cott *et al.*, 1994). Another example concerns the interaction of antigenic peptides bound to MHC molecules with specific TCR. Single amino acid substitutions are often able to change the behaviour of a peptide from stimulatory to antagonistic activity, and these differences are due mainly to changes in the kinetic constants of the interaction (Matsui *et al.*, 1994; Lyons *et al.*, 1996).

Only kinetic studies of interactions can give insight into the mechanism of binding and reveal any conformational adaptation of the binding partners. Studies of several hapten–

antibody interactions have shown that these interactions are rarely a single-step process (Lancet and Pecht, 1976; Kranz *et al.*, 1982; Hoebeke *et al.*, 1987). The binding of several peptides of the tryptophan synthase β2 subunit to some Mabs was found to be sometimes compatible with a one-step reaction (Larvor *et al.*, 1991), although more often it consisted of a first rapid 'encounter' step followed by a slower adaptation step of the two reactants (Friguet *et al.*, 1989). In some cases it was found that the Mab was able to select one particular peptide conformer (Rondard *et al.*, 1997).

Procedures used in kinetic analysis with BIAcore have been described in several reviews (Karlsson, 1994; Karlsson and Roos, 1997; Myszka, 1997; Saunal *et al.*, 1997; Schuck, 1997a). Kinetic data can be analysed at different levels of sophistication such as simple qualitative ranking, estimation of apparent rate constants for comparative purposes or detailed analysis of the interaction mechanism.

The binding process in the BIAcore can be described as two successive processes:

$$A_{bulk} \underset{L_m}{\overset{L_m}{\rightleftarrows}} A_{surface} + B \underset{k_d}{\overset{k_a}{\rightleftarrows}} AB$$

where A_{bulk} is the analyte concentration in the bulk, $A_{surface}$ is the analyte concentration at the gold surface with ligand immobilized in the dextran matrix, Lm is the mass transport coefficient (Glaser, 1993), B is the ligand concentration, AB is the analyte–ligand complex concentration, k_a the association rate constant and k_d the dissociation rate constant. The first reaction describes the transport of analyte between the bulk and the dextran matrix, and the second the binding of analyte to immobilized ligand. If the first step which describes the transport of A_{bulk} to the dextran surface is much faster than association of analyte with ligand, the observed binding will be driven by the interaction constants of A and B. On the contrary, at very high ligand concentration [B], the formation of the complex AB is very fast and proportional to $[A_{surface}] \cdot [B] \cdot k_a$. In this case, the observed binding will be limited by mass transport and kinetic data for the interaction will not be obtained. Mass transfer effects also occur for the reverse reaction when analyte dissociates in the buffer flow since, at very high levels of immobilization, dissociated analyte will rebind to the ligand. Between these extremes, the binding will be determined by contributions from both mass transport and interaction kinetics. On the basis of simulated data, it has been hypothesized that an additional mass transport effect caused by the dextran matrix may be operative (Schuck, 1996; Schuck and Minton, 1996a). However, Karlsson and Fält (1997) and Parsons and Stockley (1997) have shown that the presence of the dextran matrix does not affect the binding. To minimize mass transport effects, kinetic studies should be performed at a high flow rate (>30 μl/min) and under conditions of low surface derivatization. Rebinding can also be prevented by adding a competitor ligand during the dissociation phase (Panayotou *et al.*, 1993; Rauffer-Bruyère *et al.*, 1997). The quality of the response data can be improved by using a reference surface. Binding experiments should be run for several analyte concentrations (at least five) ranging from 0.1 to 10 K_d (starting for example with analyte concentrations ranging from 5 to 500 nM). Deviations from the simple 1:1 interaction model may arise from mass transport limitation, heterogeneity of the reactants, and from the fact that binding sites are not independent because of their proximity within the dextran surface (O'Shannessy and Winzor, 1996). Careful design and optimization of the experimental parameters are required to simplify the evaluation procedure (Myszka, 1997; Schuck, 1997b).

Evaluation procedures Kinetic information can be extracted from sensorgrams using several evaluation methods such as linearization, analytical integration or numerical

Table 5.2 Integrated rate equations used for the 1:1 interaction model analysis

Association	$R = \frac{k_aCR_{max}}{k_aCn + k_d} \cdot (1 - e^{-(k_aCn+k_d)})(t - t_0)$
Dissociation	$R = R_0 \cdot e^{-k_d(t-t_0)}$

C is the concentration of the analyte; R_{max} is the maximum response level that is related to the total concentration of the immobilized ligand (Karlsson *et al.*, 1991). The dissociation rate constant k_d and the association rate constant k_a and R_{max} are evaluated by fitting the data to the corresponding equations.

integration (Morton *et al.*, 1995). The linearization method was originally proposed by Karlsson *et al.* (1991) but has major drawbacks (Karlsson and Roos, 1997), since it is valid only for a simple 1:1 interaction model (equation 5.1).

The non-linear least squares method for analysing the sensorgrams (O'Shannessy *et al.*, 1993) is now the most used method and is applicable to more complex interactions. This method proceeds in an iterative way in order to find a theoretical curve that matches closely the experimental sensorgram, on the basis of minimal residual values. The theoretical curve is calculated from equations that express the resonance signal (R) at time (t) as a function of the kinetic parameters. Different equations can be used to calculate kinetic parameters, for instance curve fitting using integrated equations (Table 5.2) and a separate analysis of the analyte injection and subsequent buffer phase (O'Shannessy *et al.*, 1993). The dissociation rate constant is first evaluated from the post-injection part of the sensorgram, where ideally only dissociation of complex occurs. This value is necessary for evaluating the association rate constant from the analyte injection part of the sensorgram, where both association and dissociation of the binding partners occur. The k_d and k_a values should be the same for different analyte concentrations, and the validity of the interpretation can be assessed by self-consistency tests (Schuck and Minton, 1996b).

Global fitting in combination with numerical integration calculations has become the method of choice for analysing simultaneously data sets obtained at different analyte concentrations. This approach gives access to a larger choice of interaction models (Table 5.3; Karlsson and Fält, 1997) and is the only one capable of accurate estimations of kinetic rate constants for complex interactions (Morton *et al.*, 1995) involving, for instance, conformational changes (Fisher *et al.*, 1994; Glaser and Hausdorf, 1996; Karlsson and Fält, 1997).

5.4.3 ***Concentration measurements***

All methods for affinity or kinetic constants measurements require a precise knowledge of the concentration of at least one binding partner. It should be stressed that the active concentration of analyte may be lower than the nominal concentration determined, for instance, by colorimetric dosage or spectrophotometry. In such cases, the affinity K and the association rate constant k_a that are measured are underestimations of the real constants. The concentration of active antibody can be determined with BIAcore under conditions of total mass transport limitation (Karlsson *et al.*, 1994). In such cases the binding rates do not reflect the intrinsic binding capacity of the two partners, but are proportional

Table 5.3 Interaction models and differential rate equations

One to one reaction

$A + B \Leftrightarrow AB$

$dR / dt = k_a \cdot C_A \cdot (R_{max} - R) - k_d \cdot R$

One to one reaction including mass transfer

$A_{bulk} \Leftrightarrow A_{surface} + B \Leftrightarrow AB$

$dC_S / dt = k_T \cdot (C_A - C_S) - k_s \cdot C_S \cdot (R_{max} - R) + k_d \cdot R$

$dR / dt = k_a \cdot C_S \cdot (R_{max} - R) - k_d \cdot R$

Competing reactions

$A + B \Leftrightarrow AB$

$C + B \Leftrightarrow CB$

$dR_1 / dt = k_{a1} \cdot C_A \cdot (R_{max} - R_1 - R_2 \cdot p) - k_{d1} \cdot R_1$

$dR_2 / dt = k_{a2} \cdot C_C \cdot (R_{max} \cdot 1/p - R_1 \cdot 1/p - R_2) - k_{d2} \cdot R_2$

Two state reaction

$A + B \Leftrightarrow AB \Leftrightarrow AB^*$

$dR_1 / dt = k_{a1} \cdot C_A \cdot (R_{max} - R_1 - R_2) - k_{d1} \cdot R_1 - k_{a2} \cdot R_1 + k_{d2} \cdot R_2$

$dR_2 / dt = k_{a2} \cdot R_1 - k_{d2} \cdot R_2$

The differential equations are expressed in response terms. R, R_1 and R_2 correspond to the concentration of the complexes formed. C_A and C_C are the concentrations of the analytes A and C. C_S, used in the mass transfer equation, is the concentration of analyte at the sensor surface. R_{max} corresponds to the total concentration of the immobilized ligand; p is introduced as a normalizing factor when analytes of different molecular weight compete for the binding (from Karlsson and Fält, 1997).

to the concentration of analyte. The method described by Karlsson *et al.* (1994) is based on a calibration curve and requires binding partners with a high association constant ($k_a > 10^5\ M^{-1}\ s^{-1}$) as well as immobilization of a large quantity of ligand. Kazemier *et al.* (1996) adapted Karlsson's method to determine the concentration of single-chain antibody fragment (scFv). A more general method which does not require a calibration curve for determining protein concentrations with the BIAcore and which can be used under partial mass transport limitation has been developed (Christensen, 1997; Richalet-Sécordel *et al.*, 1997). This method makes it possible to determine the concentration of analyte injected at different flow rates onto a chip surface containing a high concentration of immobilized ligand. The method is applicable to ligand–analyte pairs with an association rate constant as low as $10^3\ M^{-1}\ s^{-1}$ and requires only knowledge of the molecular weight and diffusion coefficient of the analyte. Even crude samples such as an extract of recombinant proteins or culture supernatants of Mabs can be analysed. The method was used to measure active concentration of Mab, Fab fragments (Richalet-Sécordel *et al.*, 1997) as well as the amount of antibodies in immune sera (Benito and Van Regenmortel, 1998).

5.5 Analysis of structure–activity relationship

The quantitative measurement of antigen–antibody interactions makes it possible to assess which residues are implicated in the recognition process, and to determine to what extent they contribute to the binding energy. It is also possible to establish if residues located outsite of the interacting areas influence the binding. The most informative approach consists of quantitatively evaluating the effects of residue modifications of both interacting surfaces (epitope and paratope) using equilibrium affinity or kinetic rate constants. By analysing the part played by each residue in complex formation, our understanding of the chemical basis of the recognition process will increase, and this may help in the design of molecules endowed with improved binding properties.

Epitope modifications

An improved understanding of specific recognition phenomena requires both a structural and a functional analysis of the interaction (Van Regenmortel, 1989). Both approaches were used for analysing the interaction of an antibody with the immunosuppresive cyclic undecapeptide cyclosporin A (CS) (Altschuh *et al.*, 1992a; Rauffer *et al.*, 1994). Five residues had been shown to be important for binding by initial ELISA competition mapping (Quesniaux *et al.*, 1987), and the crystallographic structure of the complex showed that these residues were largely involved in the contact surface (Altschuh *et al.*, 1992a; Vix *et al.*, 1993). The fine antigenic specificity of CS was analysed by determining antibody affinity constants for 19 CS analogues (Rauffer *et al.*, 1994). The goal was to evaluate the contribution of individual side chains to the binding energy. Analogues in which a single atom or groups of atoms had been removed or added were selected, avoiding bulky substitutions. It could be shown that modifications of CS residues 3, 5 and 11 drastically affected affinity, while modifications of the other residues had less effect (Figure 5.6). The theoretical contribution to the binding energy of individual CS side chains could be calculated from the crystallographic structure (Altschuh *et al.*, 1992a; Vix *et al.*, 1993). These calculations were based on the hypothesis that the binding energy was proportional to the side chain surface buried in the complex and assumed rigid interacting molecules

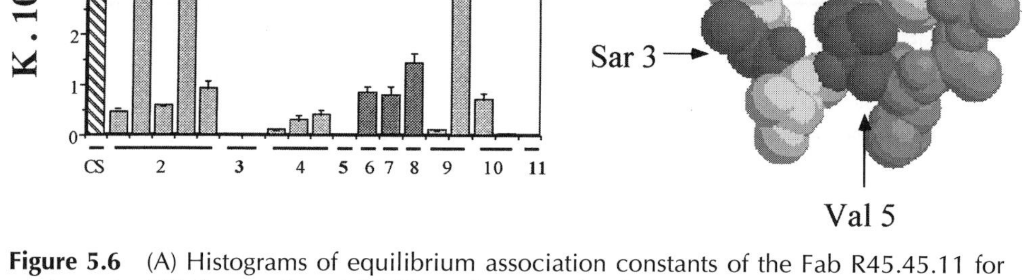

Figure 5.6 (A) Histograms of equilibrium association constants of the Fab R45.45.11 for cyclosporin (CS) (hatched) and 19 CS analogues using BIAcore. In position 2 for instance, the equilibrium association constants of five different CS analogues with altered methyl or hydroxyl groups are shown. Modification of either sarcosin 3, valine 5 or methyl valine 11 led to a drastic affinity decrease. (B) Schematic representation of the CS functional epitope recognized by the Fab R45.45.11. CS, residues (3, 5 and 11) essential for the binding of the Fab are shown in black. The seven residues whose modification leads to smaller affinity changes are shown in grey. The light grey residues (2, 4, 9, 10) are in contact with the Fab while the dark grey residues (6, 7, 8) are not.

(Chothia, 1974). For three out of six minor deletions at atomic groups, the decrease in affinity was larger than predicted from the structure. This indicated that the measured effect was complex and probably involved conformational rearrangements.

A number of other studies have confirmed that only a subset of residues in the binding site contributes to the binding affinity and energy of the complex (Alexander *et al.*, 1992; Altschuh *et al.*, 1992b; Jin *et al.*, 1992; Jin and Wells, 1994; Rondard *et al.*, 1997).

Paratope modifications

With regard to epitopes, a few hot spots of binding energy have been located in the antibody combining site (Kelley and O'Connell, 1993; Nuss *et al.*, 1993). Modifications outside the six hypervariable loops, which contribute to the structure of the paratope, have also been shown to affect antibody affinity (Chien *et al.*, 1989; Lavoie *et al.*, 1992; Hawkins *et al.*, 1993; Chatellier *et al.*, 1996a; Rauffer-Bruyère *et al.*, 1997).

Conserved residues at the V_H–V_L interface of immunoglobulins determine the interaction between variable domains of the molecule. These residues are located in the hypervariable loops as well as in framework regions of the V_H and V_L domains. Detailed comparisons of crystallographic structures of free and antigen-bound antibody have shown that antibodies may undergo rearrangements upon binding such as changes in the relative position of the V_H and V_L domains (Stanfield *et al.*, 1993). It seems that conserved residues at the V_H–V_L interface are able to influence the relative orientation of the two domains and can modulate the antigen-binding capacity of the antibody. In order to investigate their possible role in the antigen-binding properties, the effect of alanine replacements can be investigated (Chatellier *et al.*, 1996b).

Antibody–antigen recognition depends on a limited number of essential residues present at the interface between the two partners which contribute most of the binding energy. Their modification leads to a drastic decrease in affinity. In addition, other residues that are adjacent to the essential residues or which are located at a distance from the interacting surfaces may also be able to modulate the binding.

References

Alexander, H., Alexander, S., Getzoff, E.D., Tainer, J.A., Geysen, H.M. and Lerner, R.A., 1992, Altering the antigenicity of proteins, *Proceedings of the National Academy of Sciences USA*, **89**, 3352–3356.

Alonso-Whipple, C., Couet, M.L., Doss, R., Koziarz, J., Ogunro, E.A. and Crowley, W.F., 1988, Epitope mapping of human luteinizing hormone using monoclonal antibodies, *Endocrinology*, **123**, 1854–1860.

Altschuh, D., Rees, B., Vix, O. and Thierry, J.C., 1992a, Solution conformation of cyclosporin A revealed by the atomic structure of a cyclosporin–Fab complex, *Science*, **256**, 92–94.

Altschuh, D., Dubs, M.C., Weiss, E., Zeder-Lutz, G. and Van Regenmortel, M.H.V., 1992b, Determination of kinetic constants for the interaction between a monoclonal antibody and peptides using surface plasmon resonance, *Biochemistry*, **31**, 6298–6304.

Azimzadeh, A. and Van Regenmortel, M.H.V., 1991, Measurement of affinity of viral monoclonal antibodies by ELISA titration of free antibody in equilibrium mixtures, *Journal of Immunological Methods*, **141**, 199–208.

Azimzadeh, A., Pellequer, J.L. and Van Regenmortel, M.H.V., 1992, Operational aspects of antibody affinity constants measured by liquid-phase and solid-phase assays, *Journal of Molecular Recognition*, **5**, 9–18.

BENITO, A. and VAN REGENMORTEL, M.H.V., 1998, Biosensor characterization of antigenic site A of foot and mouth disease virus presented in different vector systems, *FEMS Immunology and Medical Microbiology*, **21**, 101–115.

BENJAMINI, E., 1977, Immunochemistry of the tobacco mosaic virus protein, in *Immunochemistry of Proteins*, Vol. 2, edited by M.Z. Atassi, New York: Plenum Press, pp. 265–310.

BORREBAECK, C.A.K., 1995, *Antibody Engineering*, 2nd edition, New York: Oxford University Press.

BUTLER, J.E., 1992, The behavior of antigens and antibodies immobilized on a solid phase, in *Structure of Antigens*, Vol. 1, edited by M.H.V. Van Regenmortel, Boca Raton: CRC Press, pp. 209–259.

CHATELLIER, J., RAUFFER-BRUYÈRE, N., VAN REGENMORTEL, M.H.V., WEISS, E. and ALTSCHUH, D., 1996a, Comparative interaction kinetics of two recombinant Fab fragments and of the corresponding antibodies directed to the coat protein of tobacco mosaic virus, *Journal of Molecular Recognition*, **9**, 39–51.

CHATELLIER, J., VAN REGENMORTEL, M.H.V., VERNET, T. and ALTSCHUH, D., 1996b, Functional mapping of conserved residues located at the VL and VH domain interface of a Fab, *Journal of Molecular Biology*, **264**, 1–6.

CHECOVICH, W.J., BOLGER, R.E. and BURKE T., 1995, Fluorescence polarization – a new tool for cell and molecular biology, *Nature*, **375**, 254–256.

CHIEN, N.C., ROBERTS, V.A., GUISTI, A.M., SHARF, M.D. and GETZOFF, E.D., 1989, Significant structural and functional change of an antigen-binding site by distant amino acid substitution: proposal of a structural mechanism, *Proceedings of the National Academy of Sciences USA*, **86**, 5532–5536.

CHOTHIA, C., 1974, Hydrophobic bonding and accessible surface area in proteins, *Nature*, **248**, 338–339.

CHRISTENSEN, L.H., 1997, Theoretical analysis of protein concentration determination using biosensor technology under conditions of partial mass transport limitation, *Analytical Biochemistry*, **249**, 153–164.

DAISS, J.L. and SCALICE, E.R., 1994, Epitope mapping on BIAcore: theoretical and practical considerations. *Methods: A Companion to Methods Enzymology*, **6**, 143–156.

DAISS, J.L., SCALICE, E.R. and SHARKEY, D.J., 1995, Topographical characterization of the DNA polymerase from *Thermus aquaticus*. Defining groups of inhibitor Mabs by epitope mapping function analysis using surface plasmon resonance, *Journal of Immunology*, **183**, 15–26.

DALL'ACQUA, W., GOLDMAN, E.R., EISENSTEIN, E. and MARIUZZA, R.A., 1996, A mutational analysis of the binding of two different proteins to the same antibody, *Biochemistry*, **35**, 9667–9676.

DEKKER, E.L., PORTA, C. and VAN REGENMORTEL, M.H.V., 1989, Limitations of different ELISA procedures for localizing epitopes in viral coat protein subunits, *Archives in Virology*, **105**, 269–286.

DJAVADI-OHANIANCE, L., FRIGUET, B. and GOLDBERG, M.E., 1986, Effects of ligand binding on the β2 subunit of *E. coli* tryptophan synthase analyzed with monoclonal antibodies, *Biochemistry*, **25**, 2502–2508.

DORE, I., WEISS, E., ALTSCHUH, D. and VAN REGENMORTEL, M.H.V., 1988, Visualization by electron microscopy of the location of tobacco mosaic virus epitopes reacting with monoclonal antibodies in enzyme immunoassay, *Virology*, **162**, 279–289.

DORE, I., RUHLMANN, P., OUDET, P., CAHOON, M., CASPAR, D.L.D. and VAN REGENMORTEL, M.H.V., 1990, Polarity of binding of monoclonal antibodies to tobacco mosaic virus rods and stacked disks, *Virology*, **176**, 25–29.

DOYLE, M.L., 1997, Characterization of binding interactions by isothermal titration calorimetry, *Current Opinion in Biotechnology*, **8**, 31–35.

DUBS, M.C., ALTSCHUH, D. and VAN REGENMORTEL, M.H.V., 1992, Mapping of viral epitopes with conformationally specific monoclonal antibodies using biosensor technology, *Journal of Chromatography*, **597**, 127–148.

Fägerstam, L.G., Frostell, A., Karlsson, R., Kullman, M., Larsson, A., Malmqvist, M. and Butt, H., 1990, Detection of antigen–antibody interactions by surface plasmon resonance. Application to epitope mapping, *Journal of Molecular Recognition*, **3**, 208–214.

Fisher, R.J., Fivash, M., Casas-Finet, J., Erickson, J.W., Kondoh, A., Bladen, S.V., *et al.*, 1994, Real-time DNA binding measurements of the ETS1 recombinant oncoproteins reveal significant kinetic differences between the p42 and p51 isoforms, *Protein Science*, **2**, 257–266.

Friguet, B., Chaffotte, A.F., Djavadi-Ohaniance, L. and Goldberg, M.E., 1985, Measurements of the true affinity constant in solution of antigen–antibody complexes by enzyme-linked immunosorbent assay, *Journal of Immunological Methods*, **77**, 305–319.

Friguet, B., Djavadi-Ohaniance, L. and Goldberg, M.E., 1989, Polypeptide–antibody binding mechanism: conformational adaptation investigated by equilibrium and kinetic analysis, *Research in Immunology*, **140**, 355–376.

Geysen, H.M., Meloen, R.H. and Barteling, S.J., 1984, Use of peptide synthesis to probe viral antigens for epitopes to a resolution of a single amino acid, *Proceedings of the National Academy of Sciences USA*, **81**, 3998–4002.

Glaser, R.W., 1993, Antigen–antibody binding and mass-transport by convection and diffusion to surface: a two-dimensional computer model of binding and dissociation kinetics, *Analytical Biochemistry*, **213**, 152–161.

Glaser, R.W. and Hausdorf, G., 1996, Binding kinetics of an antibody against HIV p24 core protein measured with real-time biomolecular interaction analysis suggest a slow conformational change in antigen P24, *Journal of Immunological Methods*, **189**, 1–14.

Goldberg, M.G. and Djavadi-Ohaniance, L., 1993, Methods for measurements of antibody/antigen affinity based on ELISA and RIA, *Current Opinion in Immunology*, **5**, 278–281.

Harper, M., Lema, F., Boulot, G. and Poljak, R.J., 1987, Antigen specificity and cross-reactivity of monoclonal anti-lysozyme antibodies, *Molecular Immunology*, **24**, 97–108.

Hawkins, R.E., Russell, S.J., Baier, M. and Winter, G., 1993, The contribution of contact and non-contact residues of antibody in the affinity of binding to antigen. The interaction of mutant D1.3 antibodies with lysozyme, *Journal of Molecular Biology*, **234**, 958–964.

Hoebeke, J., Engelborghs, Y., Chamat, S. and Strosberg, A.D., 1987, The immune response toward beta-adrenergic ligands and their receptors – VII. Equilibrium and kinetic binding studies of l-alprenolol to a monoclonal anti-alprenolol antibody, *Molecular Immunology*, **24**, 621–629.

Jerne, N.K., 1960, Immunological speculations, *Annual Review of Microbiology*, **14**, 341–358.

Jin, L. and Wells, J.A., 1994, Dissecting the energetics of an antibody–antigen interface by alanine shaving and molecular grafting, *Protein Science*, **3**, 2351–2357.

Jin, L., Fendly, B.M. and Wells, J.A., 1992, High resolution functional analysis of antibody–antigen interactions, *Journal of Molecular Biology*, **226**, 851–865.

Johnsson, B., Löfas, S. and Lindqvist, G., 1991, Immobilization of proteins to a carboxymethyldextran modified gold surface for biospecific interaction analysis in surface plasmon resonance, *Analytical Biochemistry*, **198**, 268–277.

Johnsson, B., Löfas, S., Lindqvist, G., Edström, A., Müller-Hillgren, R.M. and Hansson A., 1995, Comparison of methods for immobilization to carboxymethyl dextran sensor surfaces by analysis of specific activity of monoclonal antibodies, *Journal of Molecular Recognition*, **8**, 125–131.

Karlsson, R., 1994, Real-time competitive kinetic analysis of interactions between low molecular weight ligands in solution and surface-immobilized receptors, *Analytical Biochemistry*, **221**, 142–151.

Karlsson, R. and Fält, A., 1997, Experimental design for kinetic analysis of protein–protein interactions with surface plasmon resonance biosensors, *Journal of Immunological Methods*, **200**, 121–133.

Karlsson, R. and Roos, H., 1997, Reaction kinetics, in *Principles and Practice of Immunoassay*, 2nd edition, edited by C.P. Price and D.J. Newman, London: Macmillan, pp. 101–122.

KARLSSON, R., MICHAELSSON, A. and MATTSSON, L., 1991, Kinetic analysis of monoclonal antibody–antigen interactions with a new biosensor based analytical system, *Journal of Immunological Methods*, **145**, 229–240.

KARLSSON, R., ROOS, H., FÄGERSTAM, L. and PERSSON, B., 1994, Kinetic and concentration analysis using BIA technology, *Methods: A Companion to Methods in Enzymology*, **6**, 99–110.

KAZEMIER, B., DE HAARD, H., BOENDER, P., VAN GEMEN, B. and HOOGENBOOM, H., 1996, Determination of active single chain antibody concentrations in crude periplasmic fractions, *Journal of Immunological Methods*, **194**, 201–209.

KELLEY, R.F. and O'CONNELL, M., 1993, Thermodynamic analysis of an antibody functional epitope, *Biochemistry*, **32**, 6828–6835.

KRANZ, D.M., HERRON, J.N. and VOSS, E.W.J., 1982, Mechanism of ligand binding by monoclonal anti-fluorescyl antibodies, *Journal of Biological Chemistry*, **257**, 6987–6985.

LADBURY, J.E., LEMMON, M.A., ZHOU, M., GREEN, J., BOTFIELD, M.C. and SCHLESSINGER, J., 1995, Measurement of the binding of tyrosyl phosphopeptides to SH2 domains: a reappraisal, *Proceedings of the National Academy of Sciences USA*, **92**, 3199–3203.

LAKEY, J.H. and RAGGETT, E.M., 1998, Measuring protein–protein interactions, *Current Opinion in Structural Biology*, **8**, 119–123.

LANCET, D. and PECHT, I., 1976, Kinetic evidence for hapten-induced conformational transition in immunoglobulin MOPC 460, *Proceedings of the National Academy of Sciences USA*, **73**, 3549–3553.

LANE, D.P. and STEPHEN, C.W., 1993, Epitope mapping using bacteriophage peptide libraries, *Current Opinion in Immunology*, **5**, 268–271.

LARVOR, M.P., DJAVADI-OHANIANCE, L., FRIGUET, B., BALEUX, F. and GOLDBERG, M.E., 1991, Peptide/antibody recognition: synthetic peptides derived from the *E. coli* tryptophan synthase b2 subunit interact with high affinity with an anti-b2 monoclonal antibody, *Molecular Immunology*, **28**, 523–531.

LAVOIE, T.B., DROHAN, W.N. and SMITH-GIL, S.J., 1992, Experimental analysis by site-directed mutagenesis of somatic mutation effects on affinity and fine specificity in antibodies specific for lysozyme, *Journal of Immunology*, **148**, 403–513.

LEINIKKI, P.O., LEHTINEN, M., HYOTY, H., PARKKONEN, P., KANTANEN, M.-L. and HAKULINEN, J., 1993, Synthetic peptides as diagnostic tools in virology, *Advances in Virus Research*, **42**, 103–148.

LYONS, D.S., LIEBERMAN, S.A., HAMPL, J., BONIFACE, J.J., CHIEN, Y.-H., BERG, L.J. and DAVIS, M.M., 1996, A TCR binds to antagonist ligands with lower affinities and faster dissociation rates than agonists, *Immunity*, **5**, 53–61.

MALMQVIST, M., 1993, Surface plasmon resonance for detection and measurement of antibody–antigen affinity and kinetics, *Current Opinion in Immunology*, **5**, 282–286.

MALMQVIST, M., 1996, Epitope mapping by label-free biomolecular interaction analysis, *Methods. A Companion to Methods in Enzymology*, **9**, 525–532.

MANI, J.C., BRAS, J.M., AGUT, C., PAU, B., VITA, N., FERRARA, P. and BAYOL, A., 1997, Accurate topological comparison of two recombinant human growth hormones by optical surface resonance, *Analytical Biochemistry*, **248**, 50–62.

MATSUI, K., BONIFACE, J.J., STEFFNER, P., REAY, P.A. and DAVIS, M.M., 1994, Kinetics of T-cell receptor binding to peptide/I-Ek complexes: correlation of the dissociation rate with T-cell responsiveness, *Proceedings of the National Academy of Sciences USA*, **91**, 12862–12866.

MCINERNEY, T.L., HOWLETT, G.J., GRUEN, L.C. and JACKSON, D.C., 1993, Quantitative analysis of the interaction between lysozyme and monoclonal antibody D1.3, *Molecular Immunology*, **30**, 47–54.

MORRIS, G.E., 1996, *Epitope Mapping Protocols*, edited by G.E. Morris, Totowa, NJ: Humana Press.

MORTON, T.A., MYSZKA, D.G. and CHAIKEN, I.M., 1995, Interpreting complex binding kinetics from optical biosensors: a comparison of analysis by linearization, the integrated rate equation, and numerical integration, *Analytical Biochemistry*, **227**, 176–185.

MYSZKA, D.G., 1997, Kinetic analysis of macromolecular interactions using surface plasmon resonance, *Current Opinion in Biotechnology*, **8**, 50–57.

NERI, D., MONTIGIANI, S. and KIRKHAM, P.M., 1996, Biophysical methods for the determination of antibody–antigen affinities, *Trends in Biotechnology*, **14**, 465–470.

NICE, E., LAYTON, J., FABRI, L., HELLMAN, U., ENGSTROM, A., PERSSON, B. and BURGESS, A.W., 1993, Mapping of antibody- and receptor-binding domains of granulocyte colony-stimulating factor using an optical biosensor. Comparison with enzyme-linked immunosorbent assay, *Journal of Chromatography*, **646**, 159–168.

NIEBA, L., KREBBER, A. and PLÜCKTHUN, A., 1996, Competition BIAcore for measuring true affinities: large differences from values determined from binding kinetics, *Analytical Biochemistry*, **234**, 155–165.

NUSS, J.M., BOSSART-WHITAKER, P. and AIR, G.M., 1993, Identification of critical contact residues in the NC41 epitope of a subtype N9 Influenza Virus Neuraminidase, *Proteins: Structure, Function and Genetics*, **15**, 121–132.

NYGREN, H., KAARTINEN, M. and STENBERG, M., 1986, Determination by ellipsometry of the affinity of monoclonal antibodies, *Journal of Immunological Methods*, **92**, 219–225.

O'SHANNESSY, D.J. and WINZOR, D.J., 1996, Interpretation of deviations from pseudo-first-order kinetic behavior in the characterization of ligand binding by biosensor technology, *Analytical Biochemistry*, **236**, 275–283.

O'SHANNESSY, D., BRIGHAM-BURKE, M., SONESON, K.K., HENSLEY, P. and BROOKS, I., 1993, Determination of rate and equilibrium constants for macromolecular interactions using surface plasmon resonance: use of nonlinear least squares analysis methods, *Analytical Biochemistry*, **212**, 457–468.

PADLAN, E.A., 1994, Anatomy of the antibody molecule, *Molecular Immunology*, **31**, 169–217.

PANAYOTOU, G., GISH, G., END, P., TRUONG, O., GOUT, I., DHAND, R., *et al.*, 1993, Interactions between SH2 domains and tyrosine-phosphorylated platelet-derived factor resonance is not affected by the presence of a dextran matrix, *Analytical Biochemistry*, **254**, 82–87.

PARSONS, I.D. and STOCKLEY, P.G., 1997, Quantitation of the *Escherichia coli* methionine repressor-operator interaction by surface plasmon resonance is not affected by the presence of a dextran matrix, *Analytical Biochemistry*, **254**, 82–87.

PATERSON, Y., 1992, Mapping antibody binding sites on protein antigens, *Nature*, **356**, 456–457.

PHIZICKY, E.M. and FIELDS, S., 1995, Protein–protein interactions: methods for detection and analysis, *Microbiological Review*, **59**, 94–123.

QUESNIAUX, V.F.J., TEES, R., SCHREIER, M.H., WENGER, R. and VAN REGENMORTEL, M.H.V., 1987, Fine specificity of monoclonal antibodies to cyclosporine revealed by cross-reactivity studies with a panel of cyclosporine analogues, *Molecular Immunology*, **24**, 1159–1168.

RAUFFER, N., ZEDER-LUTZ, G., WENGER, R.M., VAN REGENMORTEL, M.H.V. and ALTSCHUH, D., 1994, Structure–activity relationships for the interaction between cyclosporin A derivatives and the Fab fragment of a monoclonal antibody, *Molecular Immunology*, **31**, 913–922.

RAUFFER-BRUYÈRE, N., CHATELLIER, J., WEISS, E., VAN REGENMORTEL, M.H.V. and ALTSCHUH, D., 1997, Cooperative effects of mutations in a recombinant Fab on the kinetics of antigen binding, *Molecular Immunology*, **34**, 165–173.

RICHALET-SÉCORDEL, P.M., RAUFFER-BRUYÈRE, N., CHRISTENSEN, L.L.H., OFENLOCH-HAEHNLE, B., SEIDEL, C. and VAN REGENMORTEL, M.H.V., 1997, Concentration measurement of unpurified proteins using biosensor technology under conditions of partial mass transport limitation, *Analytical Biochemistry*, **249**, 165–173.

RONDARD, P., GOLDBERG, M.E. and BEDOUELLE, H., 1997, Mutational analysis of an antigenic peptide shows recognition in a loop conformation, *Biochemistry*, **36**, 8962–8968.

SAUNAL, H. and VAN REGENMORTEL, M.H.V., 1995a, Kinetic and functional mapping of viral epitopes using biosensor technology, *Virology*, **213**, 462–471.

SAUNAL, H. and VAN REGENMORTEL, M.H.V., 1995b, Mapping of viral conformational epitopes using biosensor measurements, *Journal of Immunological Methods*, **183**, 33–41.

SAUNAL, H., WITZ, J. and VAN REGENMORTEL, M.H.V., 1993, Inhibition of *in vitro* cotranslational dissassembly of tobacco mosaic virus by monoclonal antibodies to the viral coat protein, *Journal of General Virology*, **74**, 897–900.

SAUNAL, H., KARLSSON, R. and VAN REGENMORTEL, M.H.V., 1997, Antibody affinity measurements, in *Immunochemistry II: A practical approach*, edited by A.P. Johnstone and M.W. Turner, IRL Press, pp. 1–30.

SCHUCK, P., 1996, Kinetics of ligand binding to receptor immobilized in a polymer matrix, as detected with an evanescent wave biosensor. I. A computer simulation of the influence of mass transport, *Biophysical Journal*, **70**, 1230–1249.

SCHUCK, P., 1997a, Use of surface plasmon resonance to probe the equilibrium and dynamic aspects of interactions between biological macromolecules, *Annual Review of Biophysical and Biomolecular Structure*, **26**, 541–566.

SCHUCK, P., 1997b, Reliable determination of binding affinity and kinetics using surface plasmon resonance biosensors, *Current Opinion in Biotechology*, **8**, 498–502.

SCHUCK, P. and MINTON, A.P., 1996a, Analysis of mass transport-limited binding kinetics in evanescent wave biosensors, *Analytical Biochemistry*, **240**, 262–272.

SCHUCK, P. and MINTON, A.P., 1996b, Kinetic analysis of biosensor data: elementary tests for self-consistency, *Trends in Biochemical Sciences*, **21**, 458–460.

SMITH-GILL, S.J., LAVOIE, T.B. and MAINHART, C.R., 1984, Antigenic regions defined by monoclonal antibodies correspond to structural domains of avian lysozyme, *Journal of Immunology*, **133**, 384–393.

STANFIELD, R.L., TAKIMOTO, K.M., RINI, J.M., PROFY, A.T. and WILSON, I.A., 1993, Major antigen-induced domain rearrangements in an antibody, *Structure*, **1**, 83–93.

STENBERG, E., PERSSON, B., ROOS, H. and URBANICZKY, C., 1991, Quantitative determination of surface concentration of protein with surface plasmon resonance by using radiolabelled proteins, *Journal of Colloid Interface Sciences*, **143**, 513–526.

STEWARD, M.W. and LEW, A.M., 1985, The importance of antibody affinity in the performance of immunoassays for antibody, *Journal of Immunological Methods*, **78**, 173–190.

SZASBO, A., STOLZ, L. and GRANZOW, R., 1995, Surface plasmon resonance and its use in biomolecular interaction analysis (BIA), *Current Opinion in Structural Biology*, **5**, 699–705.

TRIFILIEFF, E., DUBS, M.C. and VAN REGENMORTEL, M.H.V., 1991, Antigenic cross-reactivity potential of synthetic peptides immobilized on polyethylene rods, *Molecular Immunology*, **28**, 889–896.

TURCATTI, G., NEMETH, K., EDGERTON, M.D., MESETH, U., TALABOT, F., PEITSCH, M., *et al.*, 1996, Probing the structure and function of the tachykinin-2 receptor through biosynthetic incorporation of fluorescent amino acids at specific sites, *Journal of Biological Chemistry*, **271**, 19991–19996.

VAN COTT, T.C., BETHKE, F.R., POLONIS, A.R., GORNY, M.K., ZOLLA-PAZNER, S., REDFIEL, R.R. and BIRX, D.L., 1994, Dissociation rate of antibody-gp120 interactions is predictive of V3-mediated neutralization of HIV-1, *Journal of Immunology*, **153**, 449–458.

VAN OSS, C.J. and VAN REGENMORTEL, M.H.V. (Eds), 1994, *Immunochemistry*, New York: Marcel Dekker, Inc.

VAN REGENMORTEL, M.H.V., 1966, Plant virus serology, *Advances in Virus Reseach*, **12**, 207–271.

VAN REGENMORTEL, M.H.V., 1986, Tobacco mosaic virus. Antigenic structure, in *The Plant Viruses. II. The Rod-Shaped Plant Viruses*, edited by M.H.V. Van Regenmortel and H. Fraenkel-Conrat, New York: Plenum Press, pp. 79–104.

VAN REGENMORTEL, M.H.V., 1989, Structural and functional approaches to the study of protein antigenicity, *Immunology Today*, **10**, 266–272.

VAN REGENMORTEL, M.H.V., 1994, Use of biosensors to characterize recombinant proteins, in *Developments in Biological Standardization, Vol. 83. Genetic Stability and Recombinant Product Consistency*, edited by F. Brown and A.S. Lubiniecki, Basel: Karger, pp. 143–153.

Van Regenmortel, M.H.V., 1996, Strategies for mapping B-cell epitopes, *Methods. A Companion to Methods in Enzymology*, **9**, 465–558.

Van Regenmortel, M.H.V., 1997a, The antigen–antibody reaction, in *Principles and Practice of Immunoassay*, 2nd edition, edited by C.P. Price and D.J. Newman, London: Macmillan, pp. 15–34.

Van Regenmortel, M.H.V., 1997b, Antibody reactions with multiepitopic antigens, in *WEIR's Handbook of Experimental Immunology. Vol. 1. Immunochemistry and Molecular Immunology*, 5th edition, edited by L.A. Herzenberg, D.M. Weir, L.A. Herzenberg and C. Blackwell, Cambridge, MA: Blackwell Science, Inc., pp. 40.1–40.15.

Van Regenmortel, M.H.V. and Azimzadeh, A., 1994, Determination of antibody affinity, in *Immunochemistry*, edited by C.J. Van Oss and M.H.V. Van Regenmortel, New York: Marcel Dekker, pp. 805–827.

Van Regenmortel, M.H.V., Altschuh, D. and Zeder-Lutz, G., 1993, Tobacco mosaic virus: a model antigen to study virus–antibody interactions, *Biochimie*, **75**, 731–739.

Van Regenmortel, M.H.V., Altschuh, D., Chatellier, J., Rauffer-Bruyère, N., Richalet-Sécordel, P.M, and Saunal, H., 1997, Uses of biosensors in the study of viral antigens, *Immunological Investigations*, **26**, 67–82.

Vix, O., Rees, B., Thierry, J.C. and Altschuh, D., 1993, Crystallographic analysis of the interaction between cyclosporin A and the Fab fragment of a monoclonal antibody, *Proteins: Structure, Function and Genetics*, **15**, 339–348.

Wang, L.F., Du Plessis, D.H., White, J.R., Hyatt, A.D. and Eaton, B.T., 1995, Use of a gene-targeted phage display random epitope library to map antigenic determinant on the bluetongue virus outer capsid protein VP5, *Journal of Immunological Methods*, **178**, 1–12.

Ward, L.D., Howlett, G.J., Hemmacher, A., Weinstock, J., Yasukawa, K., Simpson, R. and Winzor, D.J., 1995, Use of biosensor with surface plasmon resonance detection for the determination of binding constants: measurements of interleukin-6 to soluble interleukin-6 receptor, *Biochemistry*, **34**, 2901–2907.

Wiseman, T., Williston, S., Brandts, J.F. and Lin, L.N., 1989, Rapid measurement of binding constants and heats of binding using a new titration calorimeter, *Analytical Biochemistry*, **179**, 131–137.

Zeder-Lutz, G., Wenger, R., Van Regenmortel, M.H.V. and Altschuh, D., 1993, Interaction of cyclosporin A with an Fab fragment or cyclophilin: affinity measurements and time dependent changes in binding, *FEBS Letters*, **326**, 153–157.

6

Antibodies as Diagnostic Tools in Plant Pathology: Summary and Outlook

LESLEY TORRANCE

Antibody-based tests play an important role in the detection and quantification of plant pathogens, and in plant disease diagnosis. The most widely used tests are enzyme immunoassays (EIA) (Tijssen, 1993) because they readily detect small amounts of pathogen in extracts of naturally infected host tissues. Also, the development of equipment to facilitate sample extraction, and to automate sample dispensing and rinsing of microplates has increased the applications of EIA in agriculture and horticulture. EIA are particularly useful where large numbers of samples must be tested annually to monitor plant health such as in vegetatively propagated crops (potatoes, small fruit, tree fruit, grapevine, and ornamental bulbs), or in tests on true seed, germplasm collections, or tissue cultures during the production of micropropagated plants. For example, in The Netherlands between 3.5 and 5 million EIA are done every year (depending on virus and aphid incidence) by the Dutch General Inspection Service to detect the potato viruses leaf roll, Y, X, S and A in seed potatoes (van den Bovenkamp, personal communication). The development of novel tests and test kits such as the dip stick, or lateral flow assay, and immunosensor technology (Ghindilis *et al.*, 1998) have broadened the applicability of antibody-based tests for example, to mycotoxins, agricultural chemicals, and the identification of genetic traits to verify crop identity. Further developments, particularly in immunosensors, will provide new opportunities for the use of rapid and sensitive tests.

The most important component of an EIA is, of course, the antibody preparation. The different kinds of antibodies, and current knowledge of antibody–antigen interactions have been described in the preceding chapters. Antibodies obtained from the sera of immunized animals (polyclonal antibodies) are the simplest to prepare and are usually of high affinity. However, because of impurities in the immunogen, they may contain antibodies that react with plant components which often produce non-specific reactions that invalidate the test or decrease the sensitivity. Furthermore, because of variations in the immune response of individual animals the properties of each antiserum differ, and although relatively large quantities can be obtained the supply is finite, and tests must be optimized for each new batch. On the other hand, monoclonal antibodies (Mabs) are produced *in vitro* from cultured hybridoma cells. Mabs are epitope specific and therefore display little cross-reactivity with plant materials. Since the cells can be stored at ultra-low temperatures in liquid nitrogen or −135°C freezers, the Mabs can be obtained in potentially unlimited supplies. However, some cell lines do not survive well in storage, and the costs of storage are high.

The focus of this book is on the development of recombinant antibodies and their applications in plant science including plant disease diagnosis. Most published results have been obtained with small single-chain variable fragments (scFv). These are expressed in *Escherichia coli* displayed on the surface of phage particles or as soluble molecules. scFv or Fab can be obtained by cloning the variable fragments of Mab genes. The expressed fragments may retain the binding properties of the parent hybridoma but some scFv are toxic to *E. coli* cells and are poorly expressed (see Chapter 2). An advantage of this approach is that it is possible to rescue the variable fragments of Mab genes from DNA of cells that will not grow again in culture after storage at low temperature. However, and more importantly, scFv can be obtained from naive or synthetic phage display libraries (Nissim *et al.*, 1994; Vaughan *et al.*, 1996; see Chapter 3). The potential advantages of the phage library approach are as follows:

1. The library can be stored conveniently (in the refrigerator or freezer) and used as a source of antibodies for any antigen.
2. The selection can be done quickly; it takes approximately two weeks to select and enrich for appropriate phage.
3. The technique bypasses the need to immunize animals, thereby facilitating production of scFv against weak immunogens or toxins.
4. Large amounts of scFv can be produced more cheaply than Mabs in high cell density bacterial fermenters.

Previously, it was shown that although scFv against plant viruses could be obtained from a synthetic phage library, they did not perform as well in immunoassays as polyclonal or monoclonal antibodies (Ziegler *et al.*, 1995; Harper *et al.*, 1997). It has been shown that the affinity of scFv can be improved by genetic manipulations to refine the binding pocket (Marks *et al.*, 1992; Schier *et al.*, 1996) but affinity maturation techniques are time consuming and labour intensive, and success is not always guaranteed (Casson and Manser, 1995). However, higher affinity scFv have been obtained directly by selections made from very large naive libraries (Griffiths *et al.*, 1994; Vaughan *et al.*, 1996). scFv are monovalent and an alternative approach is to take advantage of the increase in functional affinity obtained by making dimers, trimers or tetramers (Holliger *et al.*, 1993; Pack *et al.*, 1995; Plückthun and Pack, 1997). Also, they can be expressed in *E. coli* as fusions to other proteins for greater avidity or stability. By sub-cloning the scFv into vectors for expression as fusions to the C_L domain of immunoglobulin, or C_L Zip (Chapter 4), the proteins performed as well as existing antibodies in EIA for the detection of *Potato leafroll virus* (genus *Polerovirus*) and *Beet necrotic yellow vein virus* (genus *Benyvirus*) (see Chapters 3 and 4).

Desirable features of any assay are low cost and low detection limit, together with a simple and robust protocol. Also, the incorporation of standardized reagents is highly desirable. For example, the introduction of plant passports to guarantee the health status of plants transported between the different countries in Europe has revealed a need to use accredited tests to ensure a uniform standard, and to inspire confidence in the results of tests done throughout the European Union (EU). There are approximately ten companies worldwide that sell antibody preparations to detect plant pathogens. Many of these preparations are rabbit antisera that vary in origin and specificity; some tests incorporate Mabs. Also, many national testing laboratories use their own stocks of sera. There has been little work done to compare the performance of these different preparations.

The key question is whether recombinant antibodies have the specificity and affinity necessary to provide low cost alternatives to polyclonal sera or Mabs for use in standardized tests. The results reviewed in the preceeding chapters show that fusion proteins of scFv against plant viruses derived from naive libraries give equivalent or better results than existing reagents in EIA, and that tests can be devised which incorporate only recombinant antibodies. However, the biggest limitation to research and the widespread use of recombinant antibodies is the lack of ready access to high quality phage libraries. Production of libraries of antibody genes (constructed synthetically or from non-immunized donors) is a major undertaking. It may need an initiative at EU level to provide the resources to make a phage antibody library for use in the agricultural sciences. More research is needed to devise rapid and simple selection methods to enrich for high affinity scFv or Fab (Hoogenboom, 1997), and to select binders against cell surface antigens. The ability to genetically manipulate the scFv or Fab sequence is a great advantage. Profitable areas of future research are in the production of vectors for sub-cloning to construct different affinity matrices, and to incorporate different tags, reactive molecules, and reporter molecules for novel assays. Also, antibodies have been displayed on the surface of bacteria (Ståhl and Uhlén, 1997).

Recently, peptide mimotopes were obtained from a phage display peptide library and used as controls in immunoassays for geminiviruses (Ziegler *et al.*, 1998). Use of mimotopes or synthetic epitopes as alternatives to the distribution of pathogen preparations for positive controls in EIA will help to decrease the risk of dissemination of plant diseases. Therefore, the results of research to date indicate that engineered antibodies from phage display libraries can complement existing antibody preparations. Furthermore, in future, antibodies and peptides derived from phage display technology are likely to be valuable components of standardized tests for plant diseases.

References

Casson, L.P. and Manser, T., 1995, Evaluation of loss and change of specificity resulting from random mutagenesis of an antibody V_H region, *Journal of Immunology*, **155**, 5647–5654.

Ghindilis, A.L., Atanasov, P., Wilkins, M. and Wilkins, E., 1998, Immunosensors: electrochemical sensing and other engineering approaches, *Biosensors and Bioelectonics*, **13**, 113–131.

Griffiths, A.D., Williams, S.C., Hartley, O. and Tomlinson, I.M., 1994, Isolation of high affinity human antibodies directly from large synthetic repertoires, *The EMBO Journal*, **13**, 32456–32460.

Harper, K., Kerschbaumer, R., Ziegler, A., Macintosh, S.M., Cowan, G.H., Himmler, G., *et al.*, 1997, A scFv-alkaline phosphatase fusion protein which detects potato leafroll luteovirus in plant extracts by ELISA, *Journal of Virological Methods*, **63**, 237–242.

Holliger, P., Prospero, T. and Winter, G., 1993, 'Diabodies': small bivalent and bispecific antibody fragments, *Proceedings of the National Academy of Sciences USA*, **90**, 6444–6448.

Hoogenboom, H.R., 1997, Designing and optimizing library selection strategies for generating high-affinity antibodies, *Trends in Biotechnology*, **15**, 62–70.

Marks, J.D., Griffiths, A.D., Malmqvist, M., Clackson, T.P., Bye, J. and Winter, G., 1992, By-passing immunization: building high affinity human antibodies by chain shuffling, *Bio/Technology*, **10**, 779–783.

Nissim, A., Hoogenboom, H.R., Tomlinson, I.M., Flynn, G., Midgley, C., Lane, D. and Winter, G., 1994, Antibody fragments from a single pot phage display library as immunochemical reagents, *The EMBO Journal*, **13**, 692–698.

Pack, P., Mueller, K., Zahn, R. and Plückthun, A., 1995, Tetravalent miniantibodies with high avidity assembling in *Escherichia coli, Journal of Molecular Biology*, **246**, 28–34.

PLÜCKTHUN, A. and PACK, P., 1997, New protein engineering approaches to multivalent and bispecific antibody fragments, *Immunotechnology*, **3**, 83–105.

SCHIER, R., BALINT, R.F., McCALL, A., APELL, G., LARRICK, J.W. and MARKS, J.D., 1996, Identification of functional and structural amino-acid residues by parsimonious mutagenesis, *Gene*, **169**, 147–155.

STÅHL, S. and UHLÉN, M., 1997, Bacterial surface display: trends and progress, *Trends in Biotechnology*, **15**, 185–192.

TIJSSEN, P., 1993, Practice and theory of enzyme immunoassays, in *Laboratory Techniques in Biochemistry and Molecular Biology*, Vol. 15, eds R.H. Burdon and P.H. van Knippenberg, Elsevier, Amsterdam.

VAUGHAN, T.J., WILLIAMS, A.J., PRITCHARD, K., OSBOURN, J.K., POPE, A.R., EARNSHAW, J.C., *et al.*, 1996, Human antibodies with sub-nanomolar affinities isolated from a large non-immunized phage display library, *Nature Biotechnology*, **14**, 309–314.

ZIEGLER, A., TORRANCE, L., MACINTOSH, S.M., COWAN, G.H., and MAYO, M.A., 1995, Cucumber mosaic cucumovirus antibodies from a synthetic phage display library, *Virology*, **214**, 235–238.

ZIEGLER, A., MAYO, M.A. and TORRANCE, L., 1998, Synthetic antigen from a peptide library can be an effective positive control in immunoassays for the detection of two geminiviruses, *Phytopathology*, **88**, 1302–1305.

PART TWO

Antibodies in Plant Biotechnology

7

Production of Antibodies in Transgenic Plants – A General Introduction

CHRIS DE WILDE, GEERT DE JAEGER, MYRIAM DE NEVE,
MARC VAN MONTAGU AND ANN DEPICKER

7.1 Introduction

Transgenic plants are capable of synthesizing several types of recombinant proteins, including antibody fragments. The specific recognition and binding of virtually any type of antigen is a unique feature that causes antibodies to be exploited in numerous industrial, medical, and scientific applications. Figure 7.1 shows the different types of potential applications of antibodies produced in plants. Nowadays, commercial interest focuses on the ability of plants to produce antibodies or antibody fragments in bulk quantities. Plants show several advantages as a large-scale production system. They can be grown easily and inexpensively in large quantities and can be harvested and processed using existing infrastructures. Isolation and purification, if necessary, allow the use of plant-made antibodies for *ex planta* applications, including diagnosis and affinity-based purification as well as therapy (Ma and Hein, 1995). Whereas whole plants can be used for antibody production on an agronomical scale, plant cells cultured in bioreactors allow the production of antibodies in a contained environment with the antibodies being secreted into the culture medium. Currently, different field crops producing antibodies for *ex planta* applications are approved by the US Food and Drug Administration and grown on an industrial scale (Gibbs, 1997). Alternatively, plants can be engineered in order to synthesize antibodies intended for *in planta* uses. These applications comprise the immunomodulation of plant antigens of protein or non-protein nature to manipulate the plant's metabolism, as well as intra- or extracellular immunization (Baltimore, 1988), to obtain resistance against plant pathogens. In the near future *in planta* immunomodulation will be used to obtain transgenic crops with altered agronomic traits.

For *in planta* as well as for *ex planta* applications, the accumulation level, proper targeting, and functionality of the plant-made antibodies will be major factors in determining the success of the strategy. In this chapter, problems and uncertainties still hampering the universal exploitation of plants synthesizing antibodies will be discussed, including stability of accumulation levels, post-translational modifications, downstream processing, localization of the plant-made antibodies, and intracellular expression of antibodies. Other important factors, such as the identification of appropriate target antigens, the isolation of antibodies that bind to appropriate epitopes of these antigens, and the cloning of corresponding antibody genes are beyond the scope of this chapter and are discussed in detail in other chapters of this book.

7.2 Accumulation levels of antibodies and antibody fragments targeted to the apoplast or endoplasmic reticulum of transgenic plants

Table 7.1 summarizes the current status of antibody production in transgenic plants, with emphasis on the expression and targeting signals used, the plant species, tissues, and organs that were studied and the accumulation levels obtained. For the efficient expression of full-size antibodies in plants, the individual immunoglobulin chains should be targeted to the endoplasmic reticulum (ER) (Hiatt *et al.*, 1989; Voss *et al.*, 1995). ER-targeting is obtained by expressing each polypeptide chain as an N-terminal fusion with a signal peptide. Plant signal peptides (Düring *et al.*, 1990; De Neve *et al.*, 1993) as well as murine leader peptides (Hiatt *et al.*, 1989; Hein *et al.*, 1991; Ma *et al.*, 1994; van Engelen *et al.*, 1994; Voss *et al.*, 1995; Baum *et al.*, 1996) or yeast propeptides (Hein *et al.*, 1991) have been used successfully to direct immunoglobulin chains to the ER. Upon ER translocation, the immunoglobulin chains are expected to assemble to a complete antibody. It has not yet been investigated whether specific plant chaperones are involved in this assembly

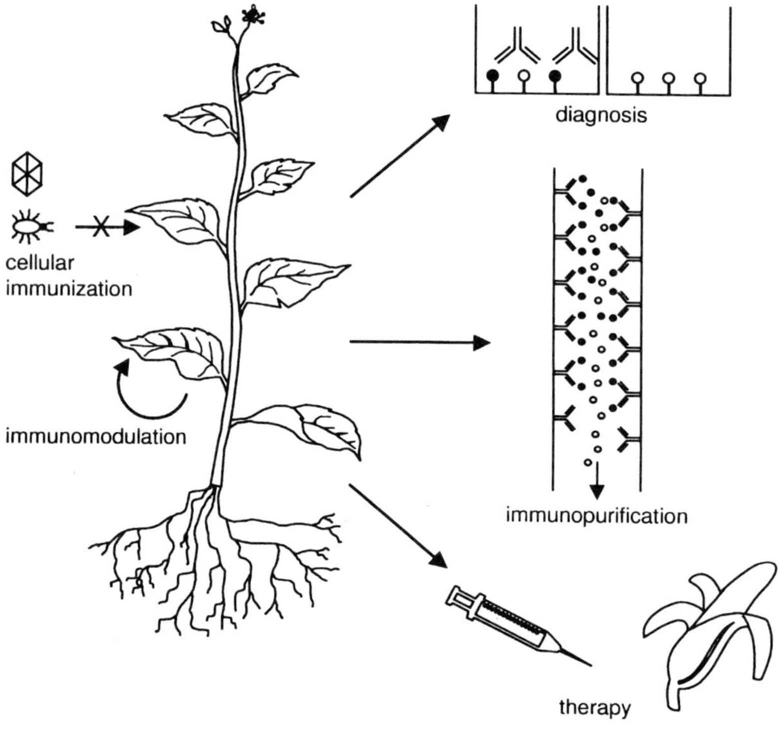

Figure 7.1 Potential applications of antibodies produced in plants. Plants can be engineered to synthesize antibodies in order to obtain pathogen resistance (cellular immunization) or to manipulate the plant's metabolism (immunomodulation). Isolation and purification, if necessary, allows the use of plant-made antibodies for diagnosis, affinity-based purification, and therapy.

Table 7.1 Current status of antibody production in transgenic plants

Type	Promoter	Location	Maximum accumulation level (per cent)[a]	Organ	Reference
IgM	P35S	Apoplast	0.01[b]	Leaf (Nt)	Baum *et al.* (1996)
			0.003[b]	Root (Nt)	
IgG2b	P35S[c]	Apoplast	0.16	Leaf (Nt)	Voss *et al.* (1995)
		Cytosol	<<	Leaf (Nt)	
IgG1	P35S	Apoplast	1.28	Leaf (Nt)	Hiatt *et al.* (1989)
			20 mg/l[d]	Callus s.c. (Nt)	Hiatt and Mostov (1993)
		Cytosol	<<	Leaf (Nt)	Hiatt *et al.* (1989)
		Apoplast	0.77	Leaf (Nt)	Hein *et al.* (1991)
IgG1	P35S	Apoplast	0.055	Leaf (Nt)	De Neve *et al.* (1993)
			1.3	Callus (Nt)	
			1.0	Callus (At)	
Fab	P35S	Apoplast	0.044	Leaf (Nt)	
			1.3	Leaf (At)	
			0.3	Callus (Nt)	
			0.8	Callus (At)	
IgG1	κ: pTR2′	Apoplast	0.6[e]	Root[g] (Nt)	van Engelen *et al.* (1994)
	γ: P35S[c]		1.10[f]	Root[g] (Nt)	
scFv	P35S[c]	Cytosol	<<	Leaf (Nt)	Schouten *et al.* (1996)
		Cytosol	0.2[h]	Leaf (Nt)	
		Apoplast	0.01	Leaf (Nt)	
		ER	1.0	Leaf (Nt)	
scFv	P35S	Cytosol	0.06	Leaf (Nt)	Owen *et al.* (1992)
		Apoplast	0.5	Leaf (Nt)	Firek *et al.* (1993)
			0.5 mg/l[d]	Callus s.c. (Nt)	
scFv	P35S	Cytosol	0.1	Leaf (Nb)	Tavladoraki *et al.* (1993)
scFv	Legumin B4	Apoplast	0.42–0.67	Seed (Nt)	Fiedler and Conrad (1995)
		Cytosol	<<	Seed (Nt)	
scFv	P35S	ER	4.8	Leaf (Nt)	Artsaenko *et al.* (1995)
		Apoplast	<<	Leaf (Nt)	
	Legumin B4	ER	4.0	Seed (Nt)	Phillips *et al.* (1997)
scFv	P35S	Apoplast	0.0055	Leaf (Nt)	Bruyns *et al.* (1996)
		Cytosol	0.0025	Leaf (Nt)	
scFv	P35S	Apoplast	0.002–0.01	Leaf (Nb)	Fecker *et al.* (1996)
			0.05–0.1	Leaf (Nb)	
scFv	P35S[c]	Cytosol	0.03	Root (St)	Schouten *et al.* (1997)
		Cytosol	0.15–0.3[h]	Root (St)	
		Cytosol	0.3[i]	Root (St)	
scFv1	P35S	ER	3.0	Leaf (Ph)	De Jaeger *et al.* (1999)
			3.0	Petal (Ph)	
		Cytosol	<<	Leaf (Ph)	
scFv2	P35S	ER	1.0	Leaf (Ph)	
			2.0	Petal (Ph)	
		Cytosol	<<	Leaf (Ph)	
			1.0	Petal (Ph)	

Table 7.1 (*cont'd*)

Type	Promoter	Location	Maximum accumulation level (per cent)[a]	Organ	Reference
scFv3	P35S	Cytosol	0.3	Leaf (Ph)	
			0.1	Petal (Ph)	
scFv4	P35S	Cytosol	<<	Leaf (Ph)	
			0.05	Petal (Ph)	
scFv5	P35S	Cytosol	<<	Leaf (Ph)	
			0.3	Petal (Ph)	
V_H	P35S	Apoplast	1.0	Leaf (Nb)	Benvenuto *et al.* (1991)

All full-size antibody-encoding genes and antibody fragment-encoding genes were cloned from monoclonal hybridoma cell lines, except for scFv1, scFv2, scFv3, scFv4, and scFv5 (De Jaeger *et al.*, 1999) that were obtained from immunized mice by phage display.
[a] The percentage antibody or antibody fragment protein of the total amount of soluble extracted protein.
[b] This value, representing the amount of monomeric IgM proteins, is an underestimation, because the corresponding pentameric IgM molecules from the hybridoma cell line were used as standard.
[c] This 35S promoter contains a double enhancer.
[d] The maximum accumulation level per litre of medium.
[e] The genes encoding heavy and light chains are in tandem orientation on the T-DNA.
[f] The genes encoding heavy and light chains are in divergent orientation on the T-DNA.
[g] Stems, leaves, and flowers from plants with a high antibody expression in roots were also found to contain antibodies with levels up to 0.4 per cent.
[h] The scFv protein contained a C-terminal KDEL peptide (Lys–Asp–Glu–Leu).
[i] The scFv protein contained a C-terminal KDEI peptide (Lys–Asp–Glu–Ile).
Abbreviations: $\ll$, below detection level; P35S, 35S promoter of the cauliflower mosaic virus; s.c., suspension culture; pRT2′, promoter of the T-DNA gene *2*; At, *Arabidopsis thaliana*; Nt, *Nicotiana tabacum*; Nb, *Nicotiana benthamiana*; Ph, *Petunia hybrida*; St, *Solanum tuberosum*; κ, kappa light chain; γ, gamma heavy chain.

process. In the absence of additional targeting or retention signals and provided that correct processing occurs, assembled antibodies or antibody fragments are expected to migrate via the secretory pathway to the cell surface, where they are secreted (for a review on protein targeting, see Kermode, 1996).

For two different IgG antibodies synthesized in tobacco, accumulation levels of approximately 1 per cent of the total soluble protein (TSP) were found (Hiatt *et al.*, 1989; van Engelen *et al.*, 1994). On the contrary, for two other IgG antibodies, much lower accumulation levels were detected in tobacco (De Neve *et al.*, 1993; Voss *et al.*, 1995), whereas the same antibody can accumulate to levels up to 3 per cent of the TSP in *Arabidopsis* (De Wilde, De Neve and Depicker, unpublished results); this observation shows that the maximum accumulation level can differ greatly, depending on the antibody that is used and the plant species that is studied. Several reports show that extracted plant-made antibodies can bind their antigen *in vitro* with the same binding characteristics as the original hybridoma-derived antibody (Hiatt *et al.*, 1989; Hein *et al.*, 1991; Ma *et al.*, 1994, 1995). This antigen-binding capacity is not the result of refolding during or after the extraction as suggested by the *in vivo* effect of plant-made antibodies. Voss *et al.* (1995) showed

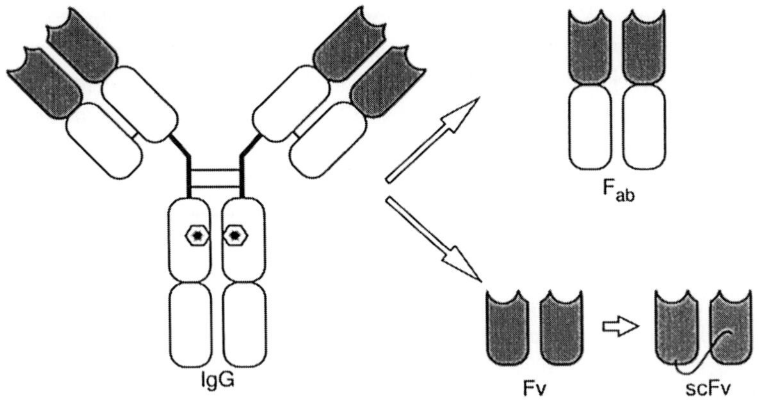

Figure 7.2 Schematic representation of the structure of a full-size IgG antibody and of a derived Fab, Fv and scFv fragment.

that in transgenic tobacco plants the amount of secreted IgG antibody directed against the capsid protein of the tobacco mosaic virus (TMV) correlated with a reduction in TMV infectivity.

Besides full-size antibodies, antibody fragments can be expressed in plants. Figure 7.2 shows the structure of a full-size IgG antibody and of a derived Fab, Fv, and scFv fragment. To date, only one report dealing with Fab expression has been published. Together with the data on the expression of a full-size secreted IgG antibody, De Neve *et al.* (1993) showed that in tobacco analogous accumulation levels can be obtained with the derived Fab fragment. In contrast, in *Arabidopsis thaliana*, tenfold higher accumulation levels were observed for the Fab than for the full-size antibody, indicating that the maximum accumulation level of antibodies or antibody fragments greatly depends on the plant species. Fab fragments extracted from plants were shown to be able to bind their antigen *in vitro*.

Most research on antibody fragment expression in plants has focused on scFv fragments. The accumulation of scFv proteins has been investigated in three plant compartments: the apoplast, the ER, and, in regard to *in planta* applications, also the cytosol (Table 7.1). For an overview of cytosolic scFv expression, we refer to the last section of this chapter. Large variations can be observed in the maximum accumulation levels of different scFv fragments targeted to the apoplast. Whereas some scFv fragments accumulate to levels of 0.5 per cent of the TSP (Firek *et al.*, 1993; Fiedler and Conrad, 1995), many other scFv fragments accumulate to low (less than 0.1 per cent) (Bruyns *et al.*, 1996; Fecker *et al.*, 1996; Schouten *et al.*, 1996) or even undetectable levels (Artsaenko *et al.*, 1995). Since normal mRNA levels were detected, the low accumulation level of certain scFv fragments is due to instability or inefficient synthesis or folding of the proteins. The highest accumulation levels (in the range of 1 per cent or even higher) have been found when the scFv proteins were targeted to and retained in the ER using an N-terminal signal peptide and a C-terminal ER retention signal KDEL (Artsaenko *et al.*, 1995; Schouten *et al.*, 1996; Phillips *et al.*, 1997; De Jaeger *et al.*, 1999). Both for secreted (Firek *et al.*, 1993; Fiedler and Conrad, 1995; Bruyns *et al.*, 1996; Fecker *et al.*, 1996) and ER-retained (Schouten *et al.*, 1996; De Jaeger *et al.*, 1999) scFv fragments, the scFv proteins could bind their antigen after extraction. One ER-retained scFv fragment was shown to be able to induce an *in vivo* effect (Artsaenko *et al.*, 1995; Phillips *et al.*, 1997). Expression of whole antibodies and antibody fragments has also been performed in callus tissue (De Neve *et al.*, 1993), callus suspension cultures (Firek *et al.*, 1993; Hiatt and Mostov, 1993), and hairy root cultures (Wongsamuth and Doran, 1997). In all these studies, the antibody or antibody fragment was provided with a signal sequence for secretion. De Neve *et al.* (1993) found IgG and Fab accumulation levels in the range of 1 per cent both in tobacco and in *Arabidopsis* callus tissue. In the medium of callus suspension cultures, Hiatt and Mostov (1993) and Firek *et al.* (1993) found 20 mg/l IgG and 0.5 mg/l scFv protein, respectively. In hairy root cultures, an IgG antibody reached accumulation levels of 18 mg/l or 1.8 per cent of TSP, of which up to 43 per cent was secreted into the medium (Wongsamuth and Doran, 1997). Finally, storage organs could be a suitable alternative to leaves for producing antibodies. Fiedler and Conrad (1995) reported the expression of an scFv antibody fragment in seeds of transgenic tobacco plants at levels of up to 0.67 per cent of the TSP; these seeds could be stored at room temperature for at least one year without any detectable loss of scFv content or activity (see Chapter 8).

From these results it is clear that plants and plant cell or tissue cultures offer many opportunities for the large-scale production of antibodies and antibody fragments. Although the maximum accumulation level depends on the plant species, the intrinsic properties of the antibody itself, and also on the tissue (De Neve *et al.*, 1993; van Engelen *et al.*, 1994), we may conclude that full-size antibodies and Fab fragments can accumulate to high levels when secreted. On the contrary, for scFv fragments it seems that the highest levels are obtained when they are retained in the ER. Unfortunately, no reports have been published dealing with ER retention of full-size antibodies or Fab fragments, nor with expression of antibody fragments in any of the organelles.

7.3 Stability of antibody accumulation levels

Antibody expression levels of 1 per cent or more of the TSP make plants an economically competitive production system. However, a problem that could hamper the general economic exploitation of plants as antibody production systems is the frequently observed gene silencing phenomenon that leads to instability of antibody accumulation levels in subsequent generations (for reviews on silencing, see Meyer and Saedler, 1996; Depicker and Van Montagu, 1997). In this respect, homozygous progeny plants, originating from a primary transformant with high antibody accumulation levels (1 per cent of the TSP) can show up to 100-fold reduced accumulation levels, suggesting that the transgenes are being silenced (De Neve *et al.*, 1999). This type of gene silencing, being established in subsequent generations of transgenic lines, depends on the gene dosage and on homology between the transgenes that are present. In general, homology between a chimeric transgene intended for antibody expression and endogenes is limited, given that expression signals are used that are not present in plants. Thus, for antibody fragments encoded by a single gene construct, such as scFv fragments, silencing of antibody genes could be avoided by growing only hemizygous transgenic plants. The use of a hemizygous seed stock will, in addition, allow the agro-biotech industry to recover continuously the investments made during development of transgenic lines. For the expression of full-size antibodies or antibody fragments made up of more than one type of polypeptide chain, the use of identical expression signals (such as the promoter or the 3′ untranslated region) to direct high expression of the different polypeptides should be carefully evaluated. Although homozygosity can result in increased antibody accumulation levels (Voss *et al.*, 1995; Fecker *et al.*, 1996), it will also enhance the chance of homology-dependent gene silencing being established in transgenes with suboptimal positioning of the control elements (De Neve *et al.*, 1999). Another important factor is the complexity of the transgene locus; selection of plant lines with simple T-DNA integration patterns (Meyer, 1996) will reduce the chance of silencing phenomena. In this respect, transformation methods yielding the least complex transgene integration patterns are preferred. Finally, the further elucidation of the parameters and the mechanisms involved in gene silencing in plants will allow the design of gene constructs and the selection of transgenic lines with low susceptibility to transgene silencing.

7.4 Glycosylation of secreted antibodies

Of all post-translational modifications, glycosylation could have the highest impact on the properties of secreted plant-made antibodies (Geisow, 1992). The carbohydrate moieties of glycoproteins can play important roles in protein folding and assembly (Helenius, 1994); this observation is particularly true for IgGs because interaction between the CH_2 domains occurs exclusively through the carbohydrate groups. Proteins bearing unusual carbohydrate modifications can be rapidly cleared from the body fluid (Dwek, 1995). Furthermore, certain carbohydrate structures have been found to be antigenic and regulatory authorities are demanding increasingly sophisticated carbohydrate analyses as part of the product validation procedure (Jenkins *et al.*, 1996).

Oligosaccharides that are attached, via an N-linked glycosidic bond, to an Asn residue within the consensus sequence Asn–X–Ser/Thr, in which X can be any amino acid except Pro or Asp, are termed N-linked glycans. Immunoglobulins of the IgG isotype contain such glycans attached to the CH_2 domain; other isotypes additionally contain glycans attached to the other constant domains of the heavy chain. These oligosaccharides can be classified into two groups: the high-mannose type and the complex-type glycans. Figure 7.3 shows a schematic representation of a high-mannose type and different complex-type glycans. The initial step of N-glycosylation is the transfer of a core unit of 14 saccharides ($Glc_3Man_9GlcNAc_2$), occurring in the ER (Driouich *et al.*, 1994). The sequential removal (trimming) of terminal glucose and mannose residues produces different types of high-mannose glycans, ranging from $Man_8GlcNAc_2$ to $Man_5GlcNAc_2$, with identical structures in plants as in other eukaryotic cells. In the Golgi complex, the action of different glycosidases and glycosyltransferases results in the conversion of high-mannose glycans into complex N-linked glycans. The complex-type glycans of plants differ from mammalian ones in the following ways: the lack of neuraminic acid, the presence of a xylose residue linked β-1,2 to the β-mannose of the core glycan, and the presence of a fucose residue in an α-1,3 linkage rather than an α-1,6 linkage (Lerouge and Faye, 1996). Also, a plant complex-type glycan with an arabinose residue instead of a xylose residue has been described (Priem *et al.*, 1993). A limited number of plant glycoproteins contain larger complex-type glycans of the laccase type (Takahashi *et al.*, 1986); these glycans have a biantennary structure with terminal fucosyl-N-acetyl lactosamine units.

Plant-made antibodies have been shown to possess complex-type glycans with the composition of terminal residues being distinct from those of the corresponding mammalian antibodies (Hein *et al.*, 1991). Although oligosaccharides and their modifications are necessary for proper glycoprotein folding and assembly, to stabilize folded domains and to provide solubility-enhancing polar surface groups that prevent aggregation of folding intermediates, the glycans of plant-made antibodies do not interfere with these processes. This fact has been demonstrated by the antigen-binding capacities of plant-made antibodies (Hiatt *et al.*, 1989; Düring *et al.*, 1990; De Neve *et al.*, 1993; Ma *et al.*, 1994, 1995; van Engelen *et al.*, 1994; Baum *et al.*, 1996; De Wilde *et al.*, 1996), with properties similar to

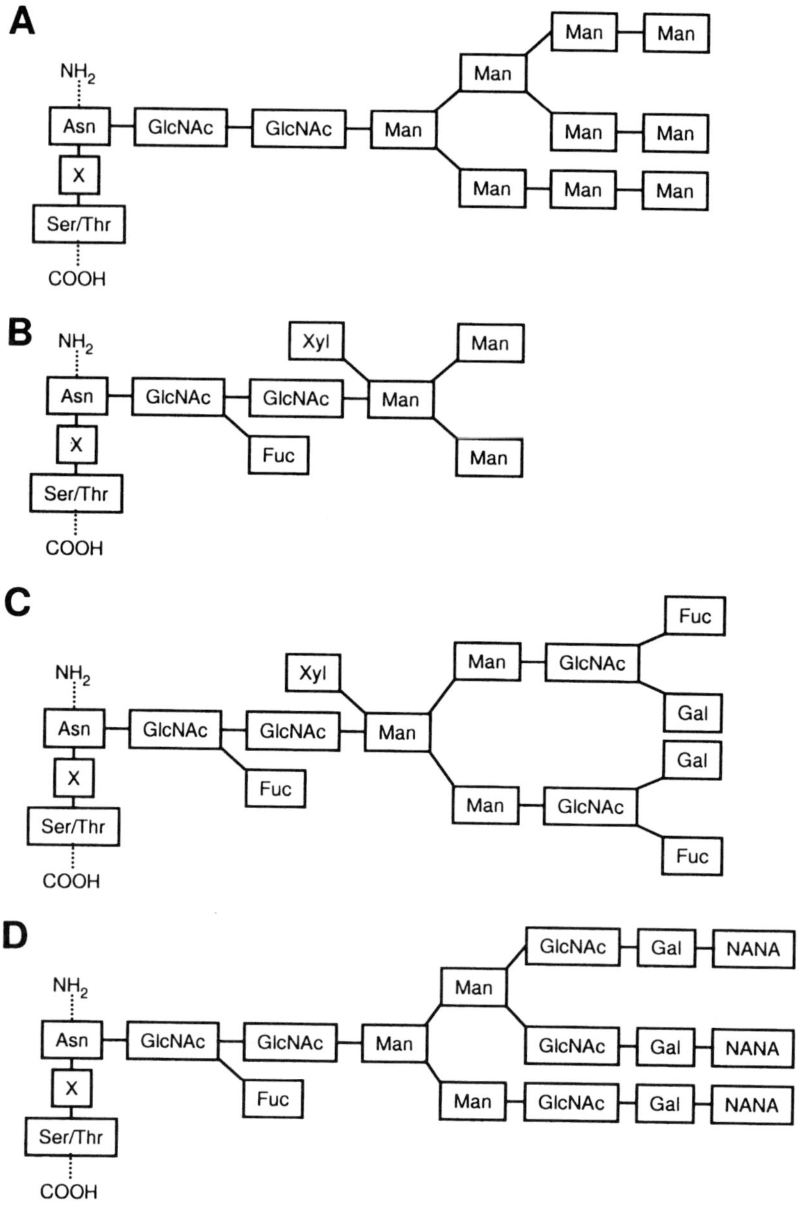

Figure 7.3 Structure of N-linked glycans of the high-mannose type and of the complex type. (A) Typical high-mannose type oligosaccharide, resulting from the removal of the terminal glucose residues from the core unit of 14 saccharides ($Glc_3Man_9GlcNc_2$); additional trimming of terminal mannose residues produces smaller glycans of the high-mannose type. These glycans are identical in plants and mammals. (B) Typical plant oligosaccharide of the complex type; most plant complex glycans are of limited size. (C) Complex-type N-linked glycan of biantennary structure having α-1,6 fucosyl-N-acetyl lactosamine units at the non-reducing end. This type of glycan has been described for a few plant glycoproteins only. (D) Typical complex-type N-linked glycan, as found on mammalian glycoproteins. Abbreviations: Asn, asparagine; Fuc, fucose; Gal, galactose: GlcNAc, *N*-acetylglucosamine; Man, mannose; NANA, *N*-acetylneuraminic acid; Ser, serine; Thr, threonine; Xyl, xylose.

those of the mammalian antibodies (Hein *et al.*, 1991; Voss *et al.*, 1995). Thus, for *in planta* applications, plant-made antibodies retain the capacities of the parental monoclonal antibodies and are able to bind the antigen in a specific manner. This observation is equally true for antibodies intended for *ex planta* applications, but plant-specific glycans could increase the immunogenicity of plant-made antibodies intended for intravascular immunotherapy. To reduce the problem of immunogenicity, the site for N-linked glycosylation could be removed from the sequence encoding the heavy chain, the proteins could be deglycosylated upon isolation, or, alternatively, plants that do not process the high-mannose glycans into complex glycans (von Schaewen *et al.*, 1993) could provide a suitable host production system. Although immunoglobulins that are obtained in this manner are not expected to be highly immunogenic, the absence of complex glycans could still cause them to be cleared from the bloodstream faster than their mammalian counterparts (Tao and Morrison, 1989; Wright and Morrison, 1994). For several applications, such as antibody-mediated radio-imaging studies, rapid clearing is one of the desired antibody features. Antibodies intended for topical immunotherapy will be less prone to recognition and degradation by the immune system, even in the absence of complex glycans of the mammalian type. Moreover, for topical immunotherapy, plant-made secretory immunoglobulins (Ma *et al.*, 1995) will most probably be advantageous as a result of their high resistance to proteolysis through the assembly of the antibody with the secretory component (Ma and Hein, 1995).

7.5 Downstream processing

Plant-made antibodies for *ex planta* applications may need downstream processing to obtain the desired level of purity. When antibodies are intended for intravascular immunotherapy, the purification demands are very high. The economics of downstream processing of antibodies from plant biomass have not been described in detail; it can be expected that the ratio of recombinant protein to total biomass will be a determining factor (Goddijn and Pen, 1995). The presence of alkaloids and other potentially toxic substances might render purification very difficult or expensive. Tobacco, until now the most commonly used host for antibody expression in plants, is a rich source of such toxic substances. In this case, the high costs of downstream processing may counteract the economic upstream production of the antibodies. A major advantage, however, of antibodies produced in transgenic plants or plant cell cultures is the absence of potentially pathogenic viruses, prions or other substances typical of animal production systems (Hemming, 1995). Therefore, expensive multistep renaturation and recovery processes are not necessary.

Plant-made antibodies, in particular secretory immunoglobulins (Ma *et al.*, 1995), have great potential as tools for passive immunization. Antibodies intended for passive immunization of mucosal surfaces of the respiratory, nasal, reproductive, or ocular systems must be purified whereas those for passive immunization of the oral cavity or the gastrointestinal tract will most probably have to be only partially purified, if at all. Important target antigens could be viruses inducing enteric diseases. It might be possible to administer the antibodies directly as an integral component of edible plant tissue. In this case, production costs of the plant-made antibodies will be very low, making it also a realistic alternative for the developing world. A major advantage of production in edible plant tissue is that antibody preparations do not need to be transported because the plant tissue can be produced locally or even domestically, making use of the existing agricultural infrastructure. Transport costs and organization are often a limiting factor in delivery of vaccines to the developing world (Moffat, 1995). A prerequisite is that plant tissue has to be eaten raw as cooking would cause the antibodies to denature and lose their activity.

Plant-made antibodies could be used for diagnostic applications. Furthermore, in analogy to the production of active industrial enzymes (Pen *et al.*, 1992, 1993; Phillipson, 1993; Aspegren *et al.*, 1995; Herbers *et al.*, 1995), plants are also suited for the production of bulk quantities of antibodies for industrial applications. Transgenic plants are a potentially competitive alternative to traditional production methods, such as microbial fermentation and animal cell cultures. Scaling-up processes are relatively simple and there is no need for expensive sterile culture media or sophisticated equipment. When an acceptable level of purity can be obtained by using a limited number of purification steps, the economics of the overall production process will most probably be advantageous. Recent results show that two or three inexpensive purification steps, such as filtration and precipitation, are sufficient to obtain 50 per cent pure antibodies from plant extracts whereas at least onc additional purification step is needed to obtain antibodies of high purity (>95 per cent) (Russel, 1997; Drossard *et al.*, 1998). Wet milling processes used in the food and feed industries have been proposed to be compatible with the separation of high-value proteins in the liquid fraction of the process (Sijmons *et al.*, 1990). Alternatively, plant-made IgGs could be purified by affinity purification, such as protein A adsorption. Although this may seem a costly process, the resulting antibody fraction can be expected to be of high purity. Moreover, the low turnover of the used reagents will reduce the overall costs. A similar purification step could be performed when antibodies are synthesized as tagged proteins (Nygren *et al.*, 1994). An interesting feature of secreted plant-made antibodies is their high abundance in intercellular fluid (ICF), namely 10–13 per cent of total soluble ICF proteins (Voss *et al.*, 1995; De Wilde *et al.*, 1996). Isolation of ICF results in functional plant-made antibodies (De Wilde *et al.*, 1996) with the number and amount of contaminating proteins being limited. In the case of cell or hairy root cultures, isolation of proteins from the medium results in a pre-purified fraction of secreted antibodies.

7.6 Accumulation sites of plant-made antibodies

The fact that after signal peptide-mediated ER targeting plant-made antibodies are indeed secreted is demonstrated by their presence in the medium of protoplast and cell suspension cultures (Hein *et al.*, 1991; van Engelen *et al.*, 1994) or in the ICF isolated from leaf tissue (Voss *et al.*, 1995; De Wilde *et al.*, 1996). The site in the plant where secreted antibodies finally accumulate will depend partly on their intrinsic properties. In the absence of chemical interactions with extracellular proteins, glycoproteins, or glycans, the size of the antibody could be the determining factor. Whereas immunolocalization of individual components suggested that secretory immunoglobulins with a molecular mass of 470 kDa accumulate in the apoplastic space between the cell wall and the plasma membrane (Ma *et al.*, 1995), full-size IgGs and Fab fragments with a molecular mass of 146 and 48 kDa, respectively, are translocated through the plant cell wall and accumulate within intercellular spaces of leaf tissues (De Wilde *et al.*, 1996). Recent data suggest that IgGs as well as Fab fragments could be transported farther in the leaf, resulting in their predominant accumulation at the interior side of the epidermal cell layers (De Wilde *et al.*, 1998). Furthermore, in roots plant-made antibodies are translocated through the cell wall and accumulate in intercellular spaces of cortex cells (De Wilde *et al.*, 1998). If these data can be extrapolated, secreted scFv fragments (Firek *et al.*, 1993; Fiedler and Conrad

1995; Bruyns *et al.*, 1996; Fecker *et al.*, 1996; Schouten *et al.*, 1996) are expected to cross the cell wall *in planta*. These localizations are important for approaches to obtain resistance against pathogens through the secretion of plant-made antibodies, because the final accumulation sites largely determine which antigens will be available as target molecules.

In the case of antibodies targeted to other compartments, it is also important to determine their exact subcellular localization in order to ascertain the possible interaction with the antigen. Several groups have synthesized scFv proteins with both an N-terminal ER-targeting signal and a C-terminal ER retention signal (Artsaenko *et al.*, 1995; Rosso *et al.*, 1996; Schouten *et al.*, 1996; De Jaeger *et al.*, 1999) resulting, as expected, in scFv accumulation in the ER (Artsaenko *et al.*, 1995; Schouten *et al.*, 1997) and in the nuclear envelope (Artsaenko *et al.*, 1995), a compartment known to be connected with the ER. The accumulation of scFv proteins that are produced without targeting signals in the cytoplasm of transgenic plants has only recently been demonstrated (Schouten *et al.*, 1997).

7.7 Immunomodulation and the problems of cytosolic expression of antibodies in plants

When plant-made antibodies are intended for immunomodulation of antigens present in the cytosol, it is a prerequisite that they are stable and functional in this compartment. Correct assembly of the individual chains of a full-size antibody or a Fab fragment is generally assumed to be impossible in the cytosol because of the lack of chaperones (Sitia and Cattaneo, 1995) and/or the reducing environment of this cell compartment that prevents the formation of disulphide bridges (Gilbert, 1990). Therefore, cytosolic expression of functional, full-size antibodies or Fab fragments seems improbable. ScFv antibody fragments were postulated to have a higher chance of being correctly folded in the cytosol (Hiatt *et al.*, 1989) because the two domains that make up each scFv are interconnected by a flexible peptide linker, avoiding the need for the assembly of different chains and the formation of inter-chain disulphide bridges. However, scFv fragments still have two intra-chain disulphide bridges that are essential for correct folding (Glockshuber *et al.*, 1992). This is believed to be the reason why only few scFv fragments accumulate even to moderate levels (0.1 per cent of the TSP) in the cytosol of plants (Owen *et al.*, 1992; Tavladoraki *et al.*, 1993). In addition, often numerous transformed plants have to be screened in order to find a plant with a moderate scFv accumulation level (Owen *et al.*, 1992). Several scFv fragments do not accumulate to detectable levels in the cytosol, even when the corresponding mRNA levels are normal (Fiedler and Conrad, 1995; Bruyns *et al.*, 1996; Schouten *et al.*, 1996), showing that the low protein accumulation levels result from instability or inefficient synthesis or folding of the proteins. Apparently, some scFv fragments seem to be stabilized at the protein level by the addition of the C-terminal peptide (DI)KDEL or KDEI with elevated accumulation levels in the cytosol as a result (Rosso *et al.*, 1996; Schouten *et al.*, 1996, 1997). It can be hypothesized that addition of a C-terminal peptide protects sterically a region of the scFv protein that is susceptible to proteolysis, or that the peptide simply prevents C-terminal protein degradation, or that it raises the stability of the protein by enhancing the number of amino acid interactions. Unfortunately, addition of a KDEL or KDEI tetrapeptide does not improve the cytosolic accumulation level of every scFv protein (Schouten *et al.*, 1997).

Besides accumulation, another prerequisite for the application of antibody expression in the cytosol is the functionality of the antibody fragment. Although for two scFv fragments a phenotypic effect was shown to be induced upon expression in the cytosol, indicating that these scFv fragments accumulated as antigen-binding proteins (Owen *et al.*, 1992; Tavladoraki *et al.*, 1993), not every scFv fragment with a high accumulation capacity in the cytosol is necessarily functional. In some cases, the absence of proper disulphide bond formation could induce conformational changes that inactivate the scFv protein or lower its affinity. Again, it seems probable that intrinsic properties of the scFv protein will determine these characteristics. Therefore, it may be advisable to search for scFv scaffolds that are stable in the cytosol without disulphide bridges being formed. Using phage display, De Jaeger *et al.* (1999) isolated five scFv fragments; one of these accumulates to high levels in the cytosol, reaching 1 per cent of TSP (Table 7.1), albeit without disulphide bonds. Moreover, this scFv protein without disulphide bonds has an affinity similar to that of the bacterially produced scFv that contains disulphide bonds. Using such a scaffold, engineering of the complementarity-determining regions would allow the immediate selection of antigen-specific scFv fragments that are suited for cytoplasmic immunomodulation.

Acknowledgements

The authors thank Koen Peeters and Tim Van Hauwermeiren for critical reading of the manuscript, Martine De Cock for excellent help preparing it, and Rebecca Verbanck for the figures. This work was supported by grants from the European Commission (FAIR CT95-0905 and BIOTECH BIO4-96-0253) and additional co-financing from the Flemish Community. C.D.W. and G.D.J. are indebted to the Vlaams Instituut voor de Bevordering van het Wetenschappelijk-Technologisch Onderzoek in de Industrie for predoctoral fellowships.

References

Artsaenko, O., Peisker, M., zur Nieden, U., Fiedler, U., Weiler, E.W., Müntz, K. and Conrad, U., 1995, Expression of a single-chain Fv antibody against abscisic acid creates a wilty phenotype in transgenic tobacco, *The Plant Journal*, **8**, 745–750.

Aspegren, K., Mannonen, L., Ritala, A., Puupponen-Pimiä, R., Kurtén, U., Salmenkallio-Marttila, M., *et al.*, 1995, Secretion of a heat-stable fungal β-glucanase from transgenic, suspension-cultured barley cells. *Molecular Breeding*, **1**, 91–99.

Baltimore, D., 1988, Intracellular immunization, *Nature* (London), **335**, 395–396.

Baum, T.J., Hiatt, A., Parrott, W.A., Pratt, L.H. and Hussey, R.S., 1996, Expression in tobacco of a functional monoclonal antibody specific to stylet secretions of the root-knot nematode, *Molecular Plant–Microbe Interactions*, **9**, 382–387.

Benvenuto, E., Ordàs, R.J., Tavazza, R., Ancora, G., Biocca, S., Cattaneo, A. and Galeffi, P., 1991, 'Phytoantibodies': a general vector for the expression of immunoglobulin domains in transgenic plants, *Plant Molecular Biology*, **17**, 865–874.

Bruyns, A.-M., De Jaeger, G., De Neve, M., De Wilde, C., Van Montagu, M. and Depicker, A., 1996, Bacterial and plant-produced scFv proteins have similar antigen-binding properties, *FEBS Letters*, **386**, 5–10.

De Jaeger, G., Buys, E., Eeckhout, D., De Wilde, C., Jacobs, A., Kapila, J., *et al.*, 1998, High level accumulation of single-chain variable fragments in the cytosol of transgenic *Petunia hybrida*, *Planta*, *European Journal of Biochemistry*, **259**, 426–434.

De Neve, M., De Loose, M., Jacobs, A., Van Houdt, H., Kaluza, B., Weidle, U., *et al.*, 1993, Assembly of an antibody and its derived antibody fragment in *Nicotiana* and *Arabidopsis*, *Transgenic Research*, **2**, 227–237.

De Neve, M., De Buck, S., De Wilde, C., Van Houdt, H., Strobbe, I., Jacobs, A., Van Montagu, M. and Depicker, A., 1998, Gene silencing results in instability of antibody production in transgenic plants. *Molecular and General Genetics*, **260**, 582–592.

DEPICKER, A. and VAN MONTAGU, M., 1997, Post-transcriptional gene silencing in plants, *Current Opinion in Cell Biology*, **9**, 373–382.

DE WILDE, C., DE NEVE, M., DE RYCKE, R., BRUYNS, A.-M., DE JAEGER, G., VAN MONTAGU, M., *et al.*, 1996, Intact antigen-binding MAK33 antibody and F_{ab} fragment accumulate in intercellular spaces of *Arabidopsis thaliana*, *Plant Science*, **114**, 233–241.

DE WILDE, C., DE RYCKE, R., BEECKMAN, T., DE NEVE, M., VAN MONTAGU, M., ENGLER, G. and DEPICKER, A., 1998, Accumulation pattern of IgG antibodies and F_{ab} fragments in transgenic *Arabidopsis thaliana* plants, *Plant Cell Physiology*, **39**, 639–646.

DRIOUICH, A., LEVY, S., STAEHELIN, L.A. and FAYE, L., 1994, Structural and functional organization of the Golgi apparatus in plant cells, *Plant Physiology and Biochemistry*, **32**, 731–749.

DROSSARD, J., LIAO, Y.-C. and FISCHER, R., 1998, Production of recombinant antibodies in plant suspension cultures, in *Recombinant Proteins from Plants: Production and Isolation of Clinically Useful Compounds*, Methods in Biotechnology, Vol. 3, C. Cunningham and A. Porter (Eds), Totowa, NJ, Humana Press, pp. 143–154.

DÜRING, K., HIPPE, S., KREUZALER, F. and SCHELL, J., 1990, Synthesis and self-assembly of a functional monoclonal antibody in transgenic *Nicotiana tabacum*, *Plant Molecular Biology*, **15**, 281–293.

DWEK, R.A., 1995, Glycobiology: towards understanding the function of sugars, *Biochemical Society Transactions*, **23**, 1–25.

FECKER, L.F., KAUFMANN, A., COMMANDEUR, U., COMMANDEUR, J., KOENIG, R. and BURGERMEISTER, W., 1996, Expression of single-chain antibody fragments (scFv) specific for beet necrotic yellow vein virus coat protein or 25 kDa protein in *Escherichia coli* and *Nicotiana benthamiana*, *Plant Molecular Biology*, **32**, 979–986.

FIEDLER, U. and CONRAD, U., 1995, High-level production and long-term storage of engineered antibodies in transgenic tobacco seeds, *Bio/Technology*, **13**, 1090–1093.

FIREK, S., DRAPER, J., OWEN, M.R.L., GANDECHA, A., COCKBURN, B. and WHITELAM, G.C., 1993, Secretion of a functional single-chain Fv protein in transgenic tobacco plants and cell suspension cultures, *Plant Molecular Biology*, **23**, 861–870.

GEISOW, M.J., 1992, Glycoprotein glycans – roles and controls, *Trends in Biotechnology*, **10**, 333–335.

GIBBS, W.W., 1997, Biotechnology – plantibodies. Human antibodies produced by field crops enter clinical trials, *Scientific American*, **277**, 23.

GILBERT, H.F., 1990, Molecular and cellular aspects of thiol-disulphide exchange, *Advances in Enzymology and Related Areas of Molecular Biology*, **63**, 69–172.

GLOCKSHUBER, R., SCHMIDT, T. and PLÜCKTHUN, A., 1992, The disulphide bonds in antibody variable domains: effects on stability, folding *in vitro* and functional expression in *Escherichia coli*, *Biochemistry*, **31**, 1270–1279.

GODDIJN, O.J.M. and PEN, J., 1995, Plants as bioreactors, *Trends in Biotechnology*, **13**, 379–387.

HEIN, M.B., TANG, Y., MCLEOD, D.A., JANDA, K.D. and HIATT, A., 1991, Evaluation of immunoglobulins from plant cells, *Biotechnology Progress*, **7**, 455–461.

HELENIUS, A., 1994, How N-linked oligosaccharides affect glycoprotein folding in the endoplasmic reticulum, *Molecular Biology of the Cell*, **5**, 253–265.

HEMMING, D., 1995, Molecular farming: using transgenic plants to produce novel proteins and other chemicals, *AgBiotech News and Information*, **7**, 19N–29N.

HERBERS, K., WILKE, I. and SONNEWALD, U., 1995, A thermostable xylanase from *Clostridium thermocellum* expressed at high levels in the apoplast of transgenic tobacco has no detrimental effects and is easily purified, *Bio/Technology*, **13**, 63–66.

HIATT, A., and MOSTOV, K. (1993). Assembly of multimeric proteins in plant cells: characteristics and uses of plant-derived antibodies, in *Transgenic Plants: Fundamentals and Applications*, A. Hiatt (Ed.), New York, Marcel Dekker, Inc., pp. 221–237.

HIATT, A., CAFFERKEY, R. and BOWDISH, K., 1989, Production of antibodies in transgenic plants, *Nature*, **342**, 76–78.

JENKINS, N., PAREKH, R.B. and JAMES, D.C., 1996, Getting the glycosylation right: implications for the biotechnology industry, *Nature Biotechnology*, **14**, 975–981.

KERMODE, A.R., 1996, Mechanisms of intracellular protein transport and targeting in plant cells, *Critical Reviews in Plant Sciences*, **15**, 285–423.

LEROUGE, P. and FAYE, L., 1996, Recent developments in structural analysis of N-glycans from plant glycoproteins, *Plant Physiology and Biochemistry*, **34**, 263–271.

MA, J.K.-C. and HEIN, M.B., 1995, Immunotherapeutic potential of antibodies produced in plants, *Trends in Biotechnology*, **13**, 522–527.

MA, J.K.-C., LEHNER, T., STABILA, P., FUX, C.I. and HIATT, A., 1994, Assembly of monoclonal antibodies with IgG1 and IgA heavy chain domains in transgenic tobacco plants, *European Journal of Immunology*, **24**, 131–138.

MA, J.K.-C., HIATT, A., HEIN, M., VINE, N.D., WANG, F., STABILA, P., *et al.*, 1995, Generation and assembly of secretory antibodies in plants, *Science*, **268**, 716–719.

MEYER, P., 1996. Repeat-induced gene silencing: common mechanisms in plants and fungi, *Biological Chemistry Hoppe-Seyler*, **377**, 87–95.

MEYER, P. and SAEDLER, H., 1996, Homology-dependent gene silencing in plants, *Annual Review of Plant Physiology and Plant Molecular Biology*, **47**, 23–48.

MOFFAT, A.S., 1995, Exploring transgenic plants as a new vaccine source, *Science*, **268**, 658–660.

NYGREN, P.A., STÅHL, S. and UHLÉN, M., 1994, Engineering proteins to facilitate bioprocessing, *Trends in Biotechnology*, **12**, 184–188.

OWEN, M., GANDECHA, A., COCKBURN, B. and WHITELAM, G., 1992, Synthesis of a functional anti-phytochrome single-chain F_v protein in transgenic tobacco, *Bio/Technology*, **10**, 790–794.

PEN, J., MOLENDIJK, L., QUAX, W.J., SIJMONS, P.C., VAN OOYEN, A.J.J., VAN DEN ELZEN, P.J.M., *et al.*, 1992, Production of active *Bacillus licheniformis* alpha-amylase in tobacco and its application in starch liquefaction, *Bio/Technology*, **10**, 292–296.

PEN, J., VERWOERD, T.C., VAN PARIDON, P.A., BEUDEKER, R.F., VAN DEN ELZEN, P.J.M., GEERSE, K., *et al.*, 1993, Phytase-containing transgenic seeds as a novel feed additive for improved phosphorus utilization, *Bio/Technology*, **11**, 811–814.

PHILLIPS, J., ARTSAENKO, O., FIEDLER, U., HORSTMANN, C., MOCK, H.-P., MÜNTZ, K. and CONRAD, U., 1997, Seed-specific immunomodulation of absicic acid activity induces a developmental switch, *EMBO Journal*, **16**, 4489–4496.

PHILLIPSON, B.A., 1993, Expression of a hybrid (1-3,1-4)-β-glucanase in barley protoplasts, *Plant Science*, **91**, 195–206.

PRIEM, B., GITTI, R., BUSH, C.A. and GROSS, K.C., 1993, Structure of ten free N-glycans in ripening tomato fruit, *Plant Physiology*, **102**, 445–458.

ROSSO, M.-N., SCHOUTEN, A., ROOSIEN, J., BORST-VRENSSEN, T., HUSSEY, R.S., GOMMERS, F.J., *et al.*, 1996, Expression and functional characterization of a single chain Fv antibody directed against secretions involved in plant nematode infection process, *Biochemical and Biophysical Research Communications*, **220**, 255–263.

RUSSEL, D., 1997, Protein production in transgenic plants, presentation at IBC International Conference on Mammalian and Plant-based Expression Systems, Washington, DC, September 22–23.

SCHOUTEN, A., ROOSIEN, J., VAN ENGELEN, F.A., DE JONG, G.A.M.I., BORST-VRENSSEN, A.W.M., ZILVERENTANT, J.F., *et al.*, 1996, The C-terminal KDEL sequence increases the expression level of a single chain antibody designed to be targeted to both the cytosol and the secretory pathway in transgenic tobacco, *Plant Molecular Biology*, **30**, 781–793.

SCHOUTEN, A., ROOSIEN, J., DE BOER, J.M., WILMINK, A., ROSSO, M.-N., BOSCH, D., *et al.*, 1997, Improving scFv antibody expression levels in the plant cytosol. *FEBS Letters*, **415**, 235–241.

SIJMONS, P.C., DEKKER, B.M.M., SCHRAMMEIJER, B., VERWOERD, T.C., VAN DEN ELZEN, P.J.M. and HOEKEMA, A., 1990, Production of correctly processed human serum albumin in transgenic plants, *Bio/Technology*, **8**, 217–221.

SITIA, R. and CATTANEO, A., 1995, Synthesis and assembly of antibodies in natural and artificial environments, in *The Antibodies*, Vol. 1, M. Zanetti and J.D. Capra (Eds), Luxembourg, Harwood Academic Publishers, pp. 127–168.

TAKAHASHI, N., HOTTA, T., ISHIHARA, H., MORI, M., TEJIMA, S., BLIGNY, R., *et al.*, 1986, Xylose-containing common structural unit in N-linked oligosaccharides of laccase from sycamora cells, *Biochemistry*, **25**, 388–395.

TAO, M.-H. and MORRISON, S.L., 1989, Studies of aglycosylated chimeric mouse–human IgG. Role of carbohydrate in the structure and effector functions mediated by the human IgG constant region, *Journal of Immunology*, **143**, 2595–2601.

TAVLADORAKI, P., BENVENUTO, E., TRINCA, S., DE MARTINIS, D., CATTANEO, A. and GALEFFI, P., 1993, Transgenic plants expressing a functional single-chain Fv antibody are specifically protected from virus attack, *Nature*, **366**, 469–472.

VAN ENGELEN, F.A., SCHOUTEN, A., MOLTHOFF, J.W., ROOSIEN, J., SALINAS, J., DIRKSE, W.G., *et al.*, 1994, Coordinate expression of antibody subunit genes yields high levels of functional antibodies in roots of transgenic tobacco, *Plant Molecular Biology*, **26**, 1701–1710.

VON SCHAEWEN, A., STURM, A., O'NEILL, J. and CHRISPEELS, M.J., 1993, Isolation of a mutant *Arabidopsis* plant that lacks N-acetyl glucosaminyl transferase I and is unable to synthesize Golgi-modified complex N-linked glycans, *Plant Physiology*, **102**, 1109–1118.

VOSS, A., NIERSBACH, M., HAIN, R., HIRSCH, H.J., LIAO, Y.C., KREUZALER, F. and FISCHER, R., 1995, Reduced virus infectivity in *N. tabacum* secreting a TMV-specific full-size antibody, *Molecular Breeding*, **1**, 39–50.

WONGSAMUTH, R. and DORAN, P., 1997, Production of monoclonal antibodies by tobacco hairy roots, *Biotechnology and Bioengineering*, **54**, 401–415.

WRIGHT, A. and MORRISON, S.L., 1994, Effect of altered C_H^{2-} associated carbohydrate structure on the functional properties and *in vivo* fate of chimeric mouse-human immunoglobulin G1, *Journal of Experimental Medicine*, **180**, 1087–1096.

8

Transgenic Plants – A Low Cost Production System for Recombinant Antibodies

ULRIKE FIEDLER, OLGA ARTSAENKO, JULIAN PHILLIPS AND UDO CONRAD

8.1 Introduction

Recombinant antibodies have attracted much interest in pharmaceutical applications. Efficient expression systems providing stable and highly active immunoglobulins have yet to be developed. Here the authors describe the use of transgenic plants as factories for the production of biotechnologically interesting antibodies. The chapter focuses on the production of recombinant single-chain Fv antibodies (scFv). Advantages and disadvantages of the plant expression system in comparison with conventional heterologous expression systems such as bacteria or insect cells are discussed, and published plant expression systems are reviewed.

8.2 Plants as production systems

Plants are used extensively for the production of various substances. In addition to their direct use as food and fodder, currently plants produce millions of tons of organic chemicals and polymers such as wood, cellulose, textile fibres and oils. Many crop plants are also used for producing starch, saccharose and lipids. The production of lipids and oils is of particular interest in providing renewable raw materials.

Plants offer many advantages over other production systems, such as bacteria, yeast or animals. Plant growth itself involves very low production costs because energy and nutrient supplies are simply sun, water, carbon dioxide and minerals. Additional fermentation costs such as those needed for microbial production do not arise.

Furthermore, in highly industrialized countries, agricultural areas are often in surplus and lie fallow. These areas could be conveniently used to produce high-quality, high-value products in plants. For developing countries an advantage of producing biomaterials in plants is that it is not necessary to construct factories or train employees in specialized techniques.

8.3 Genetically engineered plants

Over the last few centuries farmers and plant breeders have improved the productivity of plants by selective breeding. Better hybrids have been created simply by selection of the largest, strongest and most disease-resistant plants. The application of plant biotechnology allows the transfer of a greater variety of genetic information in a more precise and controlled manner. In contrast to traditional breeding where hundreds of genes are exchanged, modern plant biotechnology makes it possible to transfer only one, or a few desirable genes into a plant species. Genetic engineering enables the production of various specific compounds in plants. Plants are still, of course, used for food production but many improvements in quality (taste, ripening, solid content), yield and nutritional value (amino acid and oil composition) have been achieved by genetic engineering (Willmitzer and Töpfer, 1992; Somerville, 1994). In addition to food production, plants will be increasingly used for the isolation of a range of pharmaceutically interesting substances and for the production of biodegradable plastics. Plant biotechnology has also succeeded in developing crop plants resistant to viruses (Nelson *et al.*, 1988; Sanders *et al.*, 1992), herbicides (Shah *et al.*, 1986; Barry *et al.*, 1992), pests and insects (Jouanin *et al.*, 1998).

8.4 Production of recombinant proteins in plants

Another interesting field of plant genetic engineering is the production of foreign proteins in plants. Various foreign proteins have already been produced in plants including enzymes (Pen *et al.*, 1992; 1993), vaccines (Arntzen, 1997), and antibodies (Owen *et al.*, 1996). The expression of genes responsible for inhibiting, or encoding for certain enzymes can serve to clarify metabolic or signal-transduction pathways in plants (Herbers and Sonnewald, 1996). This can be utilized to alter certain pathways, leading to altered plant characteristics, influencing, for instance, starch or lipid composition. A completely new field in plant biotechnology is the production of pharmaceuticals which do not normally occur in plants. Until now production of these materials has been carried out in natural systems, recombinant bacterial systems (for example, expression in *E. coli*) or less often in yeast or insect cells. With these systems, expression levels may be low, especially in prokaryotic cells, as the folding pathway of a recombinant protein is often compromised. As a consequence, recombinant proteins frequently form inactive, insoluble aggregates (inclusion bodies) instead of native, soluble and biologically active protein (Stempfer *et al.*, 1996). If this is the case, functionally active proteins are obtained after isolating inclusion bodies followed by denaturing and renaturing processes. The yield of correctly folded proteins is often exceedingly low and can vary widely for different proteins. Renaturation protocols must be developed for every protein.

Transgenic plants will play an important role in the production of pharmaceutical proteins. Initial results from plant expression systems indicate that heterologous proteins are produced predominantly in soluble form. Examples include biologically active human interferon (INF alpha D) in *Brassica rapa* plants (de Zoeten *et al.*, 1989) and the synthesis of hepatitis B surface antigen (HBsAg) in transgenic tobacco (Mason *et al.*, 1992). Mason *et al.* showed that the recombinant, plant-produced HBsAg and human serum-derived HBsAg have similar physical properties. Therefore, transgenic plants have the potential to provide low-cost vaccine production systems.

Compared with expression systems such as *E. coli*, yeast or insect cells, plants offer a number of advantages: low primary energy input; no requirement for complex culture media, sterility or large culture vessels (bioreactors); easy upscaling of production; decreased waste (plant waste material can be composted); and no contamination with mammalian viruses or bacterial endotoxins. The last point is especially important for the production of recombinant proteins for use in therapy. Concerning the production of recombinant proteins in developing countries, equipment and knowledge of fermentation are often lacking. Agricultural experience in raising crops under different climatic conditions is the basic knowledge needed to raise genetically manipulated crops successfully. One important disadvantage of plant expression systems is the relatively long period of time needed to obtain transgenic plants that produce sufficient quantities of the specific protein.

8.5 Development of a plant expression system

If a foreign protein is going to be produced in plants, an existing expression system can be chosen or a new one has to be developed. Before developing a new system for large-scale production, a very important factor – the production costs – should be considered and optimized throughout the whole process. The final goal is the high-level production of foreign protein in a plant species which can be easily grown in the field. Furthermore, production must occur in a plant organ from which a considerable amount of material can be obtained and where harvesting, storage, and processing are simple or already established. For the selection of the plant species it is important to know in which geographic region the plants will be grown.

If the recombinant protein is to be used therapeutically, the chosen plant species should not contain toxic compounds that could lead to unwanted side-effects. The development of a new plant expression system includes the construction of a suitable expression cassette, its transfer into plant cells, regeneration of transgenic plants and finally, downstream processing to isolate the recombinant protein.

8.5.1 Construction of plant expression cassettes

For high-level expression of foreign proteins in plants the gene of interest is put under control of a strong promoter. The introduction of additional enhancers can increase the level of transcripts. The promoter of choice not only determines the amount but also the location of transcript synthesis. Promoters also show varying activities in different plant species. For ubiquitous expression in dicotyledonous plants (for example, tobacco, potato) the strong 35S promoter of *Cauliflower mosaic virus* (CaMV, genus *Caulimovirus*) (Benfey *et al.*, 1989) is most widely used. In comparison, organ-specific promoters often lead to higher levels of expression in the corresponding organs (for example, seeds, tubers) and the foreign proteins accumulate in these organs. Furthermore, plants expressing foreign genes exclusively in particular organs can be more productive under field conditions (Krebbers *et al.*, 1992). The expression cassettes should also include sequences of the 5′ non-translated region. For high-level expression, an optimal plant translation initiation environment and a plant terminator are both necessary. Furthermore, the codons of the foreign gene should be optimal for plant expression. In addition to sequences influencing transcription and translation, sequences for targeting the protein to specific cell compartments can be included. For cytoplasmic accumulation, the foreign genes are fused only with the promoter. In contrast, N-terminally fused signal peptides can facilitate the entry of foreign proteins into the plant secretory system by translocation across the endoplasmic reticulum (ER) membrane. Further transport depends on specific targeting, retention signals and transport competence (Chrispeels, 1991). A C-terminally fused ER retention signal (KDEL) can lead to the accumulation of foreign proteins in the ER (Munroe and Pelham, 1987). Newly synthesized proteins lacking such targeting or retention sequences are secreted via the bulk-flow or default pathway (Denecke *et al.*, 1990). This has been shown in transgenic tobacco cells for a number of proteins such as scFv (Firek *et al.*, 1993) and invertase (von Schaewen *et al.*, 1990). Other targeting signals can direct the proteins to vacuoles, chloroplasts or mitochondria.

8.5.2 *Production of transgenic plants*

The success of genetically improved plant production depends on the ability to grow single, modified cells into whole plants. Some plants – potatoes, tomatoes, tobacco, and oilseed rape – can be grown easily from single cells or plant tissue. To introduce foreign genes into these dicotyledonous plants, *Agrobacterium tumefaciens*-mediated gene transfer is the most popular technique (Klee *et al.*, 1987). For monocotyledonous plants such as corn, wheat, and soybean the gene transfer is more complicated. For transformation of these plants the following methods have proved successful: particle bombardment (Klein *et al.*, 1987), transformation of protoplasts (Krens *et al.*, 1982; Hain *et al.*, 1985), microinjection (Griesbach, 1983; Reich *et al.*, 1986), and electroporation (Lindsey and Jones, 1987). When the *Agrobacterium*-mediated gene transfer is used for production of transgenic plants the expression cassettes must be cloned into plant transformation vectors and the resulting plasmids are then transferred into *Agrobacterium tumefaciens* (Deblaere *et al.*, 1985). Clones harbouring the expression cassettes are used for transformation of suitable explants, for example, leaf discs (Horsch *et al.*, 1985) of the appropriate plant species. Plants are then regenerated. The integrity of the constructs in single transformants can be confirmed by Southern hybridization.

8.5.3 *Downstream processing*

After harvesting the plant material, synthesis of the foreign proteins can be investigated by Western blot analysis or by ELISA using an antibody against the recombinant protein. Storage conditions should be investigated and homogenization and extraction methods for obtaining soluble functionally active proteins have to be established. After the purification process, the functionality, purity and toxicity of the obtained recombinant protein must be evaluated.

8.6 Production of recombinant antibodies in plants

A new innovative field of plant biotechnology is the production of recombinant antibodies in transgenic plants. Several types of engineered antibodies have been designed in recent years. This chapter focuses on single-chain Fv antibodies (scFv). A single-chain Fv antibody is a synthetic antibody derivative consisting of the variable light-chain and the variable heavy-chain domains of an antibody molecule fused by a flexible peptide linker (Bird *et al.*, 1988; Huston *et al.*, 1988). The scFv is the smallest antibody fragment that can be constructed containing the complete antigen-binding site, and is formed by the V_L and V_H domains alone. These small recombinant antibodies retain full antigen-binding activity but lack specific assembly requirements. They are being used in diagnostics and therapeutics (Chester and Hawkins, 1995), as radiolabelled molecules (Leung *et al.*, 1995) and as fusions with immunotoxins (Chaudhary *et al.*, 1989, 1990; Batra *et al.*, 1990; Pai *et al.*, 1991; Francisco *et al.*, 1997), enzymes (Senter *et al.*, 1988; Stanislawski *et al.*, 1989; Bagshawe, 1995) or scFv of different specificity to form a so-called bispecific antibody (Fanger and Guyre, 1991). Today, scFv can be humanized by replacing the framework regions by those of a human antibody (Chester and Hawkins, 1995) or human scFv antibodies can be isolated from non-immunized phage display libraries (for example, Vaughan *et al.*, 1996; see Chapter 3). The use of these small scFv-antibody fragments offers several advantages: easy fusion with other proteins to form bifunctional proteins; good access to some intracellular compartments; and the lack of a glycosylation site which should confer low immunogenicity.

scFv antibodies have been successfully synthesized in plants and plant cells (for review see Conrad and Fiedler, 1994) as well as in bacteria. It has been shown that scFv expressed in plants and in the periplasm of *E. coli* cells have similar antigen-binding properties (Bruyns *et al.*, 1996). Of the variety of compounds expressed in transgenic plants, antibodies probably offer the widest range of applications.

In order to exploit transgenic plants for the large-scale production of recombinant antibodies, accumulation and stability of functionally active scFv-antibodies have to be optimized. Only those scFv which do not adversely interfere with plant metabolism are expected to be expressed well in plants. Expression systems allowing high-level accumulation as well as stable short- or long-term storage of antibodies in transgenic plant material will be advantageous if the harvested material has to be transported or stored before further processing.

Results obtained from several studies help to compare different plant expression systems concerning the highest accumulation rates and the best storage conditions for recombinant antibodies. Several groups have investigated the expression of scFv in different plants and plant organs as well as intra- and extracellular expression. The highest expression levels obtained for different scFv, plant species, and plant organs are shown in Table 8.1. scFv antibodies can be targeted to different cell compartments by using different targeting signals (Figure 8.1). Successful accumulation of functional antibodies obviously depends on proper targeting to the desired subcellular compartment combined with proper folding and sufficient stability.

8.6.1 Production of scFv in tobacco leaves

Expression levels of scFv in plant leaves vary from 0.01 per cent (Bruyns *et al.*, 1996) to 6.8 per cent (Fiedler *et al.*, 1997) of total soluble protein (TSP) (Table 8.1). For ubiquitous expression different plasmids have been constructed facilitating accumulation of the scFv proteins in the cytoplasm (Owen *et al.*, 1992; Tavladoraki *et al.*, 1993; Bruyns *et al.*, 1996), directing the proteins into the extracellular space (Firek *et al.*, 1993; Fecker *et al.*, 1996) or allowing accumulation in the endoplasmic reticulum (ER) (Artsaenko *et al.*, 1995; Schouten *et al.*, 1996; Fiedler *et al.*, 1997). Previous results revealed no or very low expression of scFv in the cytoplasm of plant cells (Owen *et al.*, 1992; Tavladoraki *et al.*, 1993; Fiedler and Conrad, 1995; Fecker *et al.*, 1996). Schouten *et al.* (1996, 1997) could improve cytoplasmic expression by C-terminal fusion of the scFv protein to the KDEL sequence without a signal peptide sequence. Constructs where scFv are fused N-terminally to a signal peptide are directed through the secretory pathway into the extracellular space. These

Table 8.1 Highest expression levels of the different scFv antibodies obtained in transgenic plants

scFv	Plant species	Plant organ	Construct	Highest expression	Reference
Anti-oxazolone	*Nicotiana tabacum*	Seed	USP-SS-scFv-KDEL	2.6 per cent TSP	Fiedler *et al.* (1997)
Anti-oxazolone	*Nicotiana tabacum*	Leaf	CaMV35S-SS-scFv-KDEL	4.0 per cent TSP	Fiedler *et al.* (1997)
Anti-abscisic acid	*Nicotiana tabacum*	Leaf	CaMV35S-SS-scFv-KDEL	6.8 per cent TSP	Fiedler *et al.* (1997)
Anti-abscisic acid	*Nicotiana tabacum*	Seed	USP-SS-scFv-KDEL	4.0 per cent TSP	Phillips *et al.* (1997)
Anti-phytochrome	*Nicotiana tabacum*	Leaf	CaMV35S-scFv	0.1 per cent TSP	Owen *et al.* (1992)
Anti-artichoke mottled crinkle virus	*Nicotiana benthamiana*	Leaf	CaMV35S-scFv	0.1 per cent TSP	Tavladoraki *et al.* (1993)
Anti-cutinase	*Nicotiana tabacum*	Leaf	CaMV35S-SS-scFv-KDEL	1 per cent TSP	Schouten *et al.* (1996)
Anti-β-1, 4-endoglucanase	*Solanum tuberosum*	Root	TR2′-1′-scFv-KDEL	0.3 per cent TSP	Schouten *et al.* (1997)
Anti-beet necrotic yellow vein virus coat protein/ 25 kDa non-structural protein	*Nicotiana benthamiana*	Leaf	CaMV35S-SS-scFv	0.1 per cent TSP	Fecker *et al.* (1996)
Anti-human creatine kinase-MM	*Nicotiana tabacum*	Leaf	CaMV35S-SS-scFv CaMV35S-scFv	0.01 per cent TSP	Bruyns *et al.* (1996)

expression systems resulted in increased expression compared with the cytoplasmic ones. A significant increase in accumulation was only obtained by retaining the scFv proteins in the lumen of the ER (Fiedler *et al.*, 1997).

In our studies, we have used two different scFv: scFv-ox that binds to the hapten oxazolone (Fiedler and Conrad, 1995), and scFv-ABA (Artsaenko *et al.*, 1994) that binds to the phytohormone abscisic acid. The influence of the site of antibody accumulation in the transgenic plant cell on accumulation level, antibody activity and antibody stability during storage was investigated in these studies. For ubiquitous expression in tobacco plants, transcription of the scFv genes was controlled by the strong CaMV 35S promoter. Entry of the scFv antibodies into the secretory pathway was achieved by introducing a signal-peptide-encoding sequence of the seed-storage-protein gene LeB4 (Bäumlein *et al.*, 1991a). For accumulation in the ER the scFv genes were fused at their 5′-ends with the LeB4-signal-peptide-encoding sequence and additionally at the 3′-ends with the KDEL-encoding sequence (Munroe and Pelham, 1987). High-level accumulation of the two

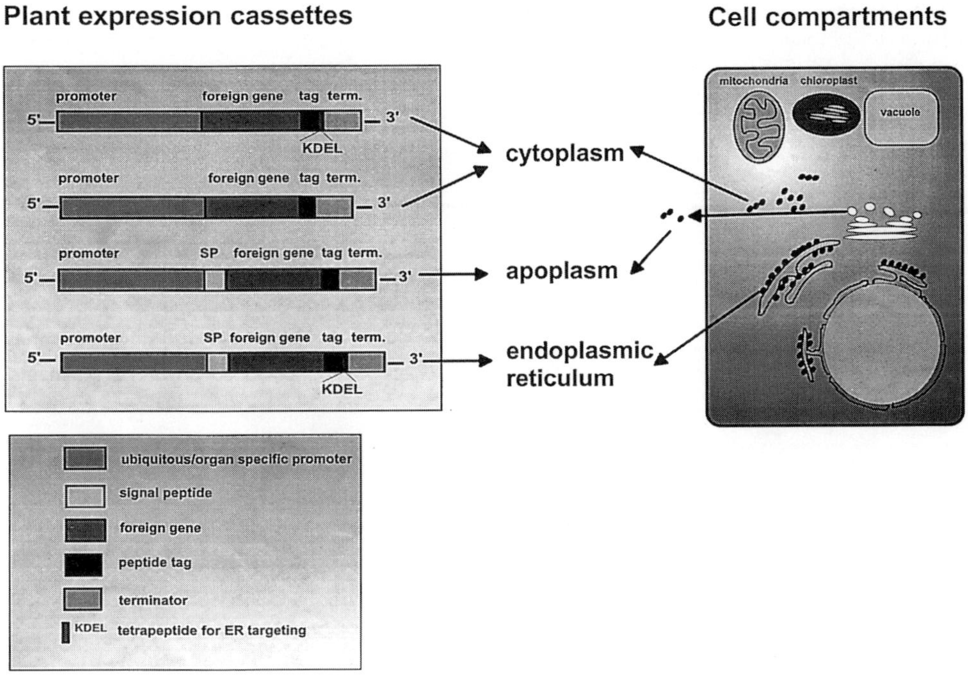

Figure 8.1 Schematic representation of cassettes for ubiquitous and organ-specific expression of scFv genes in transgenic plants (left) and corresponding plant cell compartments used for accumulation of single-chain Fv proteins (right).

different scFv proteins in transgenic tobacco plants was only achieved by retention of the recombinant proteins in the lumen of the endoplasmic reticulum. Accumulation levels and antibody properties such as stability and activity were investigated. When only a signal peptide was fused the scFv-ox and the proteins were directed to the extracellular space, scFv protein accumulation levels were relatively low – on average 0.1 per cent of leaf TSP. Retention in the ER resulted in a 10- to 20-fold increase of scFv-ox antibody concentration and also in very high accumulation of scFv-ABA (up to 6.8 per cent of TSP). Localization of the recombinant antibodies in the ER was confirmed by electron microscopy studies (Figure 8.2). The scFv proteins were detected in the endoplasmic reticulum and in the nuclear envelope.

Similar results were reported by Schouten *et al.* (1996), who obtained a 100-fold increase of scFv accumulation but only to a maximum of 1 per cent of leaf TSP in transgenic tobacco by fusion with the ER retention signal. Extended retention of the scFv proteins in the ER is thought to promote correct protein folding leading to higher scFv stability and accumulation. Accumulated scFv proteins could be extracted and affinity purified (Figure 8.3) from leaves and were able to bind to the appropriate antigen in ELISA.

We could demonstrate that retention in the ER not only leads to high expression in leaves, but also provides excellent protection of the expressed recombinant proteins in the leaf material. Transgenic tobacco leaves dried at room temperature facilitated stable storage of the ER-accumulated scFv for over three weeks. Leaf material containing the scFv-ox antibody was harvested from adult plants. In parallel, samples of leaves expressing the antibody were subsequently frozen and stored at –20°C or dried at room temperature. After drying and three weeks of further storage, proteins were extracted. Both the amount

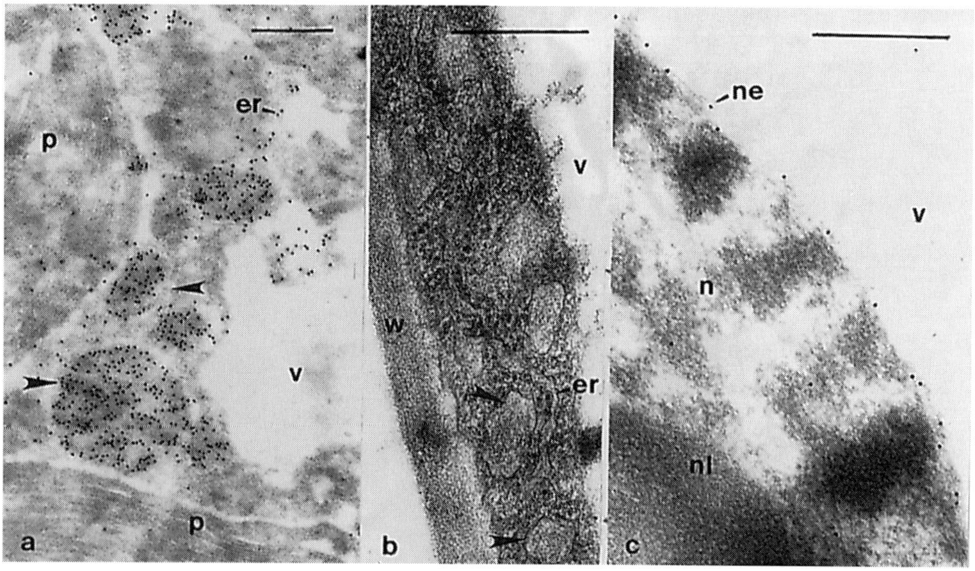

Figure 8.2 Detection of scFv-ABA proteins in transgenic tobacco leaf cells by immunogold labelling. Intracellular localization of anti-ABA scFv antibodies in the ER (a) and ER derived vesicles (arrow) and in the nuclear envelope (c) of a mesophyll cell. scFv antibodies are indicated as dark points of colloidal gold. In (b) the ultrastructure of a cytoplasmic area containing ER and ER derived vesicles is shown. Bar: 0.5 μm; er, endoplasmic reticulum; n, nucleus; ne, nuclear envelope; nl, nucleolus; p, plastid; v, vacuole; w, cell wall (micrographs are a courtesy of U. zur Nieden).

of the scFv proteins, determined by immunoblot analysis (Figure 8.4A), and their binding activity investigated by ELISA (Figure 8.4B), remained unchanged.

8.6.2 *Production of scFv in tobacco seeds*

Seeds as natural storage organs deposit proteins for long periods of time without degradation (Vandekerckhove *et al.*, 1989; Pen and Sijmons, 1993). The use of a seed-specific promoter avoids ubiquitous expression in the plant so that basic plant metabolic processes are not strongly affected. This often leads to a higher local accumulation. The harvesting process can be focused on specific plant organs. Furthermore, seeds are an optimal system for easy storage of the recombinant proteins.

During our studies, investigating the storage organ-specific accumulation of the scFv genes in transgenic tobacco, expression was controlled by two seed-specific promoters: the LeB4 (Bäumlein *et al.*, 1991a) and the USP promoter (Bäumlein *et al.*, 1991b) for *Vicia faba*. The seed-specific promoters differ in strength and temporal activity during seed development. The USP promoter was shown to be active during the early stages of seed development (Fiedler *et al.*, 1993) whereas the LeB4 promoter only becomes active during the middle to late stages (Fiedler and Conrad, 1995). Constructs for seed-specific expression were designed allowing either scFv-protein secretion into the extracellular space or retention in the ER. Previously, we have shown that the scFv-ox fused to a signal peptide accumulated up to 0.6 per cent of the TSP in transgenic tobacco seeds (Fiedler and Conrad, 1995). In this case storage, organ-specific expression was controlled

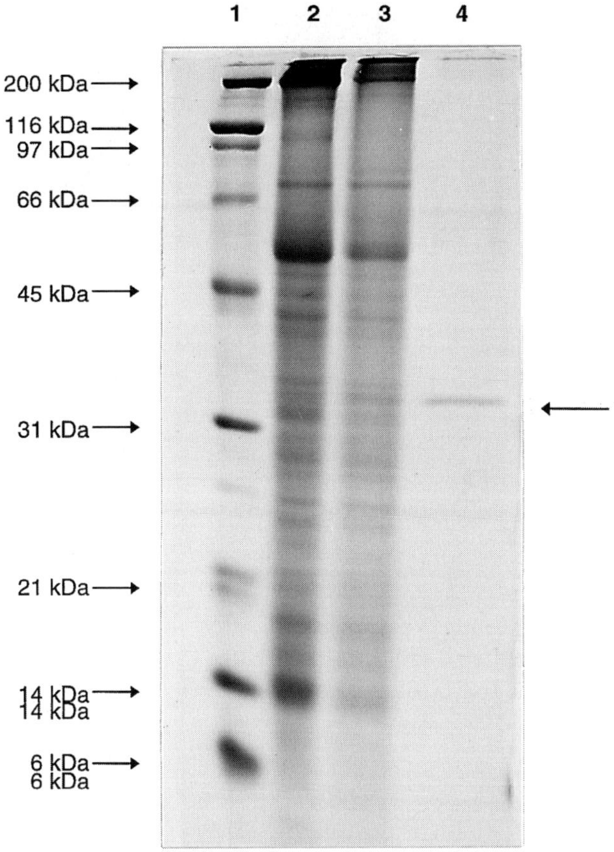

Figure 8.3 Affinity purification of scFv-ox antibodies from tobacco leaves. SDS-PAGE of leaf proteins extracted from a wild-type tobacco plant, a transgenic plant containing 1.9 per cent scFv-ox of TSP and affinity-purified scFv-ox. Lane 1, protein marker; lane 2, 30 μg TSP of a wild-type plant; lane 3, 30 μg TSP of the transgenic plant; lane 4, purified scFv-ox protein.

by the legumin B4 promoter. Electron microscopy studies revealed accumulation of these proteins in protein bodies situated near the nucleus (Figure 8.5). Significantly higher accumulation of the scFv proteins in transgenic tobacco seeds was achieved by retention of the recombinant proteins in the lumen of the ER. The highest expression levels in ripe tobacco seeds were obtained by using the USP promoter together with the ER retention signal KDEL (Phillips *et al.*, 1997). Expression levels of scFv antibodies reached up to 3–4 per cent TSP in ripe tobacco seeds. Up to 2.6 per cent scFv protein was measured in the case of scFv-ox and even higher amounts, up to 4 per cent of TSP, in the case of scFv-ABA. Under control of the USP promoter scFv proteins start to be synthesized early during seed development. If scFv proteins are retained in the ER from the beginning of synthesis the level of accumulation in seeds increases constantly. This results in higher expression controlled by the USP promoter than in LeB4 promoter plants because of extended time for synthesis and accumulation during seed development.

Mature transgenic tobacco seeds facilitate stable storage of the scFv over an extended period of time. Ripe seeds have been stored at room temperature for more than one year. The scFv protein stability and antigen-binding activity did not change during storage.

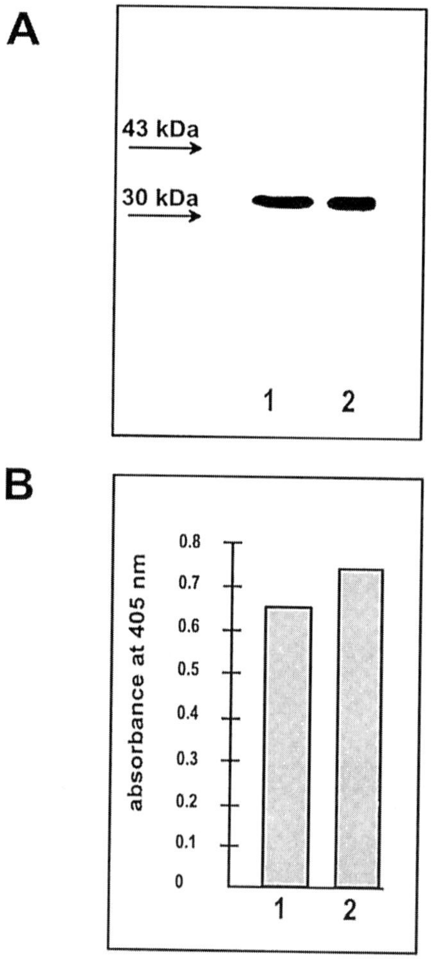

Figure 8.4 Stable storage of active scFv-ox proteins in dried, transgenic tobacco leaves shown by Western blot analysis (A) and ELISA (B). Leaves from transgenic plants expressing the scFv-ox were harvested, samples were frozen and dried at room temperature in parallel and 40 µg of extracted protein were used for Western blot analysis (A) and measurement of antigen-binding activity by ELISA (B). 1, frozen material; 2, dried material.

8.7 Conclusions

Plant genetic engineering offers the pharmaceutical industry exciting new opportunities for producing large quantities of pharmaceutical products, using much less primary energy than with bacterial, fungal or animal systems. Storage of transgenic plant material and prevention of degradation of transgenic products are important goals of the phytofarming technology. Transgenic tobacco plants have been demonstrated to tolerate high levels of scFv expression with no apparent detriment to growth or phenotype when using scFv which bind to antigens not normally found in plants. Results from different groups revealed that scFv may only accumulate to very low levels, or not at all, in the cytoplasm of plant cells (Owen *et al.*, 1992; Tavladoraki *et al.*, 1993; Fiedler and Conrad, 1995). Proteins have to be directed through the secretory pathway into the extracellular space or

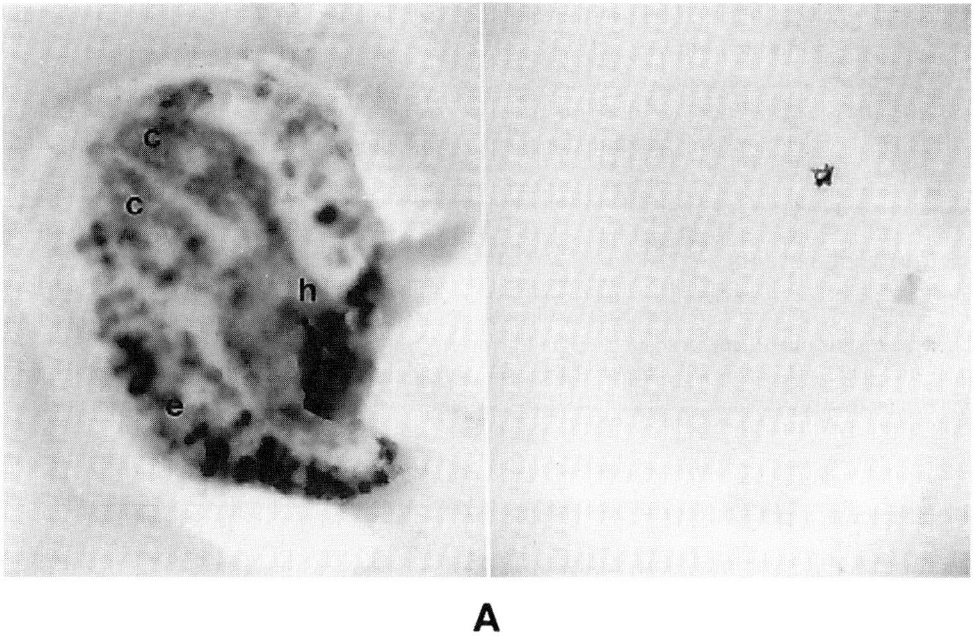

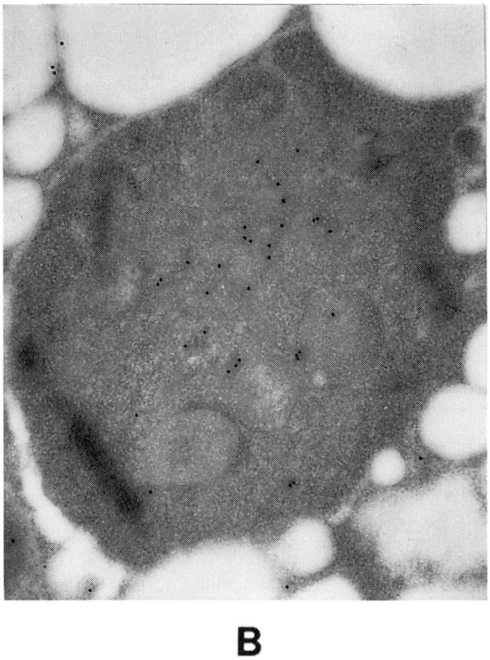

Figure 8.5 Tissue print with ripe tobacco seeds expressing scFv-ox (A) and the corresponding intracellular localization detected by immunogold labelling (B). (A) Detection of scFv proteins in mature seeds of transgenic tobacco plants by tissue printing. Left: transgenic plant expressing 0.6 per cent scFv-ox of total soluble seed protein. Right: control plant. c, cotyledon; h, hypocotyl; e, endosperm. (B) Intracellular localization of scFv-ox antibodies in mature seeds of transgenic plants. Protein bodies from the cotyledons of embryo cells are shown. scFv-ox antibodies are indicated as dark dots of colloidal gold. Magnification × 40 000.

retained in the lumen of the ER. It has been demonstrated that the accumulation levels and the stability in different organs of different scFv are similar. In summary, plant expression systems in which scFv-proteins are targeted to the ER provide the highest accumulation levels of functional scFv antibodies and allow long-term storage of the foreign protein in the harvested plant material. Therefore, it is possible to produce large quantities of immunologically active recombinant antibodies in the leaves and seeds of transgenic tobacco plants. Long-term storage of the plant material is possible without any loss of scFv or antigen-binding activity.

Transgenic plants may provide low cost, large scale, production systems when optimized plant expression systems are used to produce scFv antibodies at high levels. Obvious advantages of such systems include the ease of producing and storing functional antibody fragments.

Acknowledgements

We would like to thank I. Tillack and C. Liebing for excellent technical assistance and Dr U. zur Nieden (Institut für Pflanzenbiochemie, Halle) for electron microscopy investigations. The work described here was financially supported by the Bundesministerium für Bildung, Wissenschaft, Forschung und Technologie (BEO/210310261A) and by the Deutsche Forschungsgemeinschaft (Mu 925/1-2).

References

ARNTZEN, C.J., 1997, High-tech herbal medicine: plant-based vaccines, *Nature Biotechnology*, **15**, 221–222.

ARTSAENKO, O., WEILER, E.W., MÜNTZ, K. and CONRAD, U., 1994, Construction and functional characterisation of a single-chain Fv antibody binding to the plant hormone abscisic acid, *Journal of Plant Physiology*, **144**, 427–429.

ARTSAENKO, O., PEISKER, M., ZUR NIEDEN, U., FIEDLER, U., WEILER, E.W., MÜNTZ, K. and CONRAD, U., 1995, Expression of a single-chain Fv antibody against abscisic acid creates a wilty phenotype in transgenic tobacco, *The Plant Journal*, **8**, 745–750.

BAGSHAWE, K.D., 1995, Antibody-directed enzyme prodrug therapy – a review, *Drug Development Research*, **34**, 220–230.

BARRY, G., KISHORE, G., PADGETTE, S. and TAYLOR, M., 1992, Inhibitors of amino acid biosynthesis: strategies for imparting glyphosate tolerance to crop plants, in *Biosynthesis and Molecular Regulation of Amino Acids in Plants*, Singh, B.K., Flores, H.E. and Shannon, J.C. (Eds), Am. Soc. of Plant Phys., pp. 139–145.

BATRA, J.K., FITZGERALD, D.J., GATELY, M., CHAUDHARY, V.K. and PASTAN, I., 1990, Anti-Tac(Fv)-PE40, a single-chain antibody Pseudomonas fusion protein directed at interleukin 2 receptor bearing cells, *Journal of Biological Chemistry*, **265**, 15198–15202.

BÄUMLEIN, H., BOERJAN, W., NAGY, I., PANITZ, R., INZE, D. and WOBUS, U., 1991a, Upstream sequences regulating legumin gene expression in heterologous transgenic plants, *Molecular General Genetics*, **225**, 121–128.

BÄUMLEIN, H., BOERJAN, W., NAGY, I., BASSÜNER, R., VAN MONTAGU, M., INZE, D. and WOBUS, U., 1991b, A novel seed protein gene from *Vicia faba* is developmentally regulated in transgenic tobacco and *Arabidopsis* plants, *Molecular General Genetics*, **225**, 459–467.

BENFEY, P.N., REN, L. and CHUA, N.-H., 1989, The CaMV 35S enhancer contains at least two domains which can confer different developmental and tissue specific expression patterns, *European Journal of Molecular Biology*, **8**, 2195–2202.

BIRD, R.E., HARDMAN, K.D., JACOBSON, J.W., JOHNSON, S., KAUFMAN, B.M., LEE, S.-M., *et al.*, 1988, Single-chain antigen-binding proteins, *Science*, **242**, 423–426.

BRUYNS, A.M., DE JAEGER, G., DE NEVE, M., DE WILDE, C., VAN MONTAGU, M. and DEPICKER, A., 1996, Bacterial and plant-produced scFv proteins have similar antigen-binding properties, *FEBS Letters*, **386**, 5–10.

CHAUDHARY, V.K., QUEEN, C., JUNGHANS, R.P., WALDMANN, T.A., FITZGERALD, D.J. and PASTAN, I., 1989, A recombinant immunotoxin consisting of two antibody variable domains fused to *Pseudomonas* exotoxin, *Nature*, **339**, 394–397.

CHAUDHARY, V.K., GALLO, M.G., FITZGERALD, D.J. and PASTAN, I., 1990, A rapid method of cloning functional variable-region antibody genes in *Escherichia coli* as single-chain immunotoxins, *Proceedings of the National Academy of Sciences USA*, **87**, 9491–9494.

CHESTER, K.A. and HAWKINS, R.E., 1995, Clinical issues in antibody design, *Trends in Biotechnology*, **13**, 294–300.

CHRISPEELS, M.J., 1991, Sorting of proteins in the secretory system, *Annual Reviews in Plant Physiology of Plant Molecular Biology*, **42**, 35–49.

CONRAD, U. and FIEDLER, U., 1994, Expression of engineered antibodies in plant cells, *Plant Molecular Biology*, **26**, 1023–1030.

DEBLAERE, R., BYTEBIER, B., DE GREVE, H., DEBOECK, F., SCHELL, J., VAN MONTAGU, M. and LEEMANS, J., 1985, Efficient octopine Ti plasmid-derived vectors for *Agrobacterium*-mediated gene transfer to plants, *Nucleic Acids Research*, **13**, 4777–4788.

DENECKE, J., BOTTERMAN, J. and DEBLAERE, R., 1990, Protein secretion in plant cells can occur via a default pathway, *Plant Cell*, **2**, 51–59.

DE ZOETEN, G.A., PENSWICK, J.R., HORISBERGER, M.A., AHL, P., SCHULTZE, M. and HOHN, T., 1989, The expression, localization, and effect of a human interferon in plants, *Virology*, **172**, 213–222.

FANGER, M.W. and GUYRE, P.M., 1991, Bispecific antibodies for targeted cellular cytotoxicity, *Trends in Biotechnology*, **9**, 375–380.

FECKER, L.F., KAUFMANN, A., COMMANDEUR, U., COMMANDEUR, J., KOENIG, R. and BURGEMEISTER, W., 1996, Expression of single chain antibody fragments (scFv) specific for beet necrotic yellow vein virus coat protein or 25 kDa protein in *Escherichia coli* and *Nicotiana benthamiana*, *Plant Molecular Biology*, **32**, 979–986.

FIEDLER, U. and CONRAD, U., 1995, High-level production and long-term storage of engineered antibodies in transgenic tobacco seeds, *Bio/Technology*, **10**, 1090–1094.

FIEDLER, U., FILISTEIN, R., WOBUS, U. and BÄUMLEIN, H., 1993, A complex ensemble of cis-regulatory elements controls the expression of a *Vicia faba* non-storage seed protein gene, *Plant Molecular Biology*, **22**, 669–679.

FIEDLER, U., PHILLIPS, J., ARTSAENKO, O., MÜNTZ, K. and CONRAD, U., 1997, Optimization of scFv antibody production in transgenic plants, *Immunotechnology*, **3**, 205–216.

FIREK, S., DRAPER, J., OWEN, M.R.L., GANDECHA, A., COCKBURN, B. and WHITELAM, G., 1993, Secretion of a functional single-chain Fv protein in transgenic tobacco plants and cell suspension cultures, *Plant Molecular Biology*, **23**, 861–870.

FRANCISCO, J.A., GAWLAK, S.L. and SIEGALL, C.B., 1997, Construction, expression, and characterization of BD1-G28-5 scFv, a single-chain anti-CD40 immunotoxin containing the ribosome-inactivating protein bryodin 1, *Journal of Biological Chemistry*, **272**, 24165–24169.

GRIESBACH, R.J., 1983, Protoplast microinjection, *Plant Molecular Biology Reports*, **1**, 32–37.

HAIN, R., STABEL, P., CZERNILOFSKY, A.P., STEINBISS, H.H., HERRERA-ESTRELLA, L. and SCHELL, J., 1985, Uptake, integration, expression and genetic transmission of a selectable chimeric gene by plant protoplasts, *Molecular General Genetics*, **199**, 161–168.

HERBERS, K. and SONNEWALD, U., 1996, Manipulating metabolic partitioning in transgenic plants, *Trends in Biotechnology*, **14**, 198–205.

HORSCH, R.B., FRY, J.E., HOFFMANN, N.L., EICHHOLTZ, D., ROGERS, S.G. and FRALEY, R.T., 1985, A simple and general method for transferring genes into plants, *Science*, **227**, 1229–1231.

HUSTON, J.S., LEVINSON, D., MUDGETT-HUNTER, M., TAI, M.-S., NOVOTNY, J., MARGOLIES, M.N., *et al.*, 1988, Protein engineering of antibody binding sites: recovery of specific activity in an anti-digoxin single-chain Fv analogue produced in *Escherichia coli*, *Proceedings of the National Academy of Sciences USA*, **85**, 5879–5883.

JOUANIN, L., BONADEBOTTINO, M., GIRARD, C., MORVAT, G. and GIBAUD, M., 1998, Transgenic plants for insect resistance, *Plant Science*, **131**, 1–11.

KLEE, H., HORSCH, R. and ROGERS, S., 1987, *Agrobacterium* mediated plant transformation and its further applications to plant biology, *Annual Reviews in Plant Physiology*, **38**, 467–486.

Klein, T.M., Wolf, E.D., Wu, R. and Sandford, J.C., 1987, High-velocity microprojectiles for delivering nucleic acids into living cells, *Nature*, **327**, 70–73.

Krebbers, E., Bosch, D. and Vandekerckhove, J., 1992, Prospects and progress in the production of foreign proteins and peptides in plants, in *Plant Protein Engineering*, Shewry, P.R. and Gutteridge, S. (Eds), University Press, Cambridge, pp. 315–325.

Krens, F.A., Molendijk, L., Wullems, G.J. and Schilperoort, R.A., 1982, *In vitro* transformation of plant protoplasts with Ti-plasmid DNA, *Nature*, **296**, 72–74.

Leung, S.O., Karacay, H., Losman, M.J., Griffiths, G.L., Goldenberg, D.M. and Hansen, H.J., 1995, Bacterial expression of a kemptide fusion protein facilitates 32P labeling of a humanized, anti-carcinoembryonic antigen (hMN-14) antibody fragment, *Cancer Research*, **55**, 5968–5972.

Lindsey, K. and Jones, M.G.K., 1987, The permeability of electroporated cells and protoplasts of sugarbeet, *Planta*, **172**, 346–355.

Mason, H.S., Lam, D.M. and Arntzen, C.J., 1992, Expression of hepatitis B surface antigen in transgenic plants, *Proceedings of the National Academy of Sciences USA*, **89**, 11745–11749.

Munroe, S. and Pelham, H., 1987, A C-terminal signal prevents secretion of luminal ER proteins, *Cell*, **48**, 899–907.

Nelson, R.S., McCormick, S.M. and Delannay, X., 1988, Virus tolerance, plant growth and field performance of transgenic tomato plants expressing coat protein from tobacco mosaic virus, *Bio/Technology*, **6**, 403–409.

Owen, M.R.L., Gandecha, A., Cockburn, B. and Whitelam, G., 1992, Synthesis of a functional anti-phytochrome single-chain Fv protein in transgenic tobacco, *Bio/Technology*, **10**, 790–794.

Owen, M.R.L., Cockburn, B. and Whitelam, G., 1996, The expression of recombinant antibody fragments in plants, in *Transgenic Plants: A Production System for Industrial and Pharmaceutical Proteins*, Owen, M.R.L. and Pen, J. (Eds), pp. 245–260, John Wiley & Sons, Chichester, 1996.

Pai, L.H., Gallo, M.G., FitzGerald, D.J. and Pastan, I., 1991, Antitumor activity of a transforming growth factor alpha-Pseudomonas exotoxin fusion protein (TGF-alpha-PE40), *Cancer Research*, **51**, 2808–2812.

Pen, J. and Sijmons, P.C., 1993, Protein production in transgenic crops: analysis of plant molecular farming, in *Transgenic Plants – Fundamentals and Applications*, Hiatt, A. (Ed.), pp. 238–251, Dekker, New York.

Pen, J., Molendyk, L., Quax, W.J., Symons, P.C., van Qoijen, A.J.J., van den Elzen, P.J.M., *et al.*, 1992, Production of active *Bacillus licheniformis* alpha-amylase in tobacco and its application in starch liquefaction, *Bio/Technology*, **10**, 292–296.

Pen, J., Verwoerd, T.C., van Paridon, P.A., Beudeker, R.F., van den Elzen, P.J.M., Geerse, K., *et al.*, 1993, Phytase-containing transgenic seeds as a novel feed additive for improved phosphorus utilization, *Bio/Technology*, **11**, 811–814.

Phillips, J., Artsaenko, O., Fiedler, U., Horstmann, C., Mock, H.-P., Müntz, K. and Conrad, U., 1997, Seed-specific immunomodulation of abscisic acid activity induces a developmental switch, *European Journal of Molecular Biology*, **16**, 4489–4496.

Reich, T.J., Iyer, V.J. and Miki, B.L., 1986, Efficient transformation of alfalfa protoplasts by the intranuclear microinjection of Ti plasmids, *Bio/Technology*, **8**, 1001–1003.

Sanders, P.R., Sammons, B. and Kaniewski, H., 1992, Field resistance of transgenic tomatoes expressing the tobacco mosaic virus coat protein genes, *Phytopathology*, **82**, 683–690.

Schouten, A., Roosien, J., van Engelen, F.A., de Jong, G.A.M., Borst-Vrenssen, A.W.M., Zilverentant, J.F., *et al.*, 1996, The C-terminal KDEL sequence increases the expression level of a single-chain antibody designed to be targeted to both the cytosol and the secretory pathway in transgenic tobacco, *Plant Molecular Biology*, **30**, 781–793.

Schouten, A., Roosien, J., de Boer, J.M., Wilmink, A., Rosso, M.-N., Bosch, D., *et al.*, 1997, Improving scFv antibody expression levels in the plant cytosol, *FEBS Letters*, **415**, 235–241.

SENTER, P.D., SAUTMER, M.G., SCHREIBER, G.J., HIRSCHBERG, D.L., BROWN, J.P., HELLSTRÖM, I. and HELLSTRÖM, K.E., 1988, Anti-tumor effects of antibody-alkaline phosphatase conjugates in combination with etoposide phosphate, *Proceedings of the National Academy of Sciences USA*, **85**, 4842–4846.

SHAH, D.M., HORSCH, R.B. and KLEE, H.J., 1986, Engineering herbicide tolerance in transgenic plants, *Science*, **233**, 478–481.

SOMERVILLE, C.R., 1994, Production of industrial materials in transgenic plants, in *The Production and Uses of Genetically Transformed Plants*, Bevan, M.W., Harrison, B.D. and Leaver, C.J. (Eds), The University Press, Cambridge.

STANISLAWSKI, M., ROUSSEAU, V. and ITO, H., 1989, Immunotoxins containing glucose oxidase and lactoperoxidase with tumoricidal properties: *in vitro* killing effectiveness in a mouse plasmacytoma cell model, *Cancer Research*, **49**, 5497–5504.

STEMPFER, G., HÖLL-NEUGEBAUER, B. and RUDOLPH, R., 1996, Improved refolding of an immobilized fusion protein, *Nature Biotechnology*, **14**, 329–334.

TAVLADORAKI, P., BENVENUTO, E., TRINCA, S., MARTINIS, D.D., CATTANEO, A. and GALEFFI, P., 1993, Transgenic plants expressing a functional single-chain Fv antibody are specifically protected from virus attack, *Nature*, **366**, 469–472.

VANDEKERCKHOVE, J., VAN DAMME, J., VAN LIJSEBETTENS, M., BOTTERMAN, J., DE BLOCK, M., VANDEWIELE, M., *et al.*, 1989, Enkephalins produced in transgenic plants using modified 2S seed storage proteins, *Bio/Technology*, **7**, 929–932.

VAUGHAN, T.J., WILLIAMS, A.J., PRITCHARD, K., OSBOURN, J.K., POPE, A.R., EARNSHAW, J.C., *et al.*, 1996, Human antibodies with sub-nanomolar affinities isolated from a large non-immunized phage display library, *Nature Biotechnology*, **14**, 309–314.

VON SCHAEWEN, A., STITT, M., SCHMIDT, R., SONNEWALD, U. and WILLMITZER, L., 1990, Expression of a yeast-derived invertase in the cell wall of tobacco and Arabidopsis plants leads to accumulation of carbohydrate and inhibition of photosynthesis and strongly influences growth and phenotype of transgenic tobacco plants, *European Journal of Molecular Biology*, **9**, 3033–3044.

WILLMITZER, L. and TÖPFER, R., 1992, Manipulation of oil, starch and protein composition, *Current Opinion in Biotechnology*, **3**, 176–180.

9

Intracellular Immunomodulation in Plants – A New Tool for the Investigation of Phytohormones

OLGA ARTSAENKO, JULIAN PHILLIPS, ULRIKE FIEDLER, MARTIN PEISKER AND UDO CONRAD

9.1 Introduction

Antibody engineering and production is one of the fastest growing fields in molecular biology. The manipulation of genes through recombinant DNA technology has enabled the construction of highly specific antibodies and antibody fragments which can be expressed in a variety of host organisms. The intracellular expression of recombinant antibodies directed against specific cellular antigens ranging from regulatory factors to pathogens and their targeting into different cellular compartments can be used to block the biological functions of a selected antigen. This approach, known as intracellular immunization, has been successfully applied to inactivate oncoproteins (Biocca *et al.*, 1993, 1995) and to inhibit replication of the human immunodeficiency virus (Marasco *et al.*, 1993; Duan *et al.*, 1994) in mammalian cells. The scFv antibodies targeted into the endoplasmic reticulum (ER) have been used to functionally inactivate cell-surface receptors that are implicated in human cancer (Richardson *et al.*, 1995). The demonstration in recent years that plants are capable of producing functional whole antibodies as well as scFv antibodies offers new possibilities for modulating plant cellular antigens. The feasibility of this approach for engineering pathogen resistance was illustrated by Tavladoraki *et al.* (1993). The cytosolic expression of an scFv fragment directed against the coat protein of the plant *Artichoke mottled crinkle virus* (genus *Tombusvirus*) caused a significant reduction of viral infection and delay in symptom development in transgenic plants. Transgenic plants producing an anti-TMV (*Tobacco mosaic virus*, genus *Tobamovirus*) whole antibody showed a reduction in numbers of TMV-induced necrotic lesions which correlated with the level of antibody expression (Voss *et al.*, 1995). Immunomodulation of the activity of an endogenous photoregulatory protein, phytochrome, by means of cytosol-targeted scFv was shown by Owen *et al.* (1992). Blocking the activity of a host cellular antigen leading to a loss of function represents a powerful technique for investigating the regulatory role of a selected target. In this chapter, we illustrate with the example of abscisic acid, how the intracellular immunomodulation approach may be applied to phytohormone science.

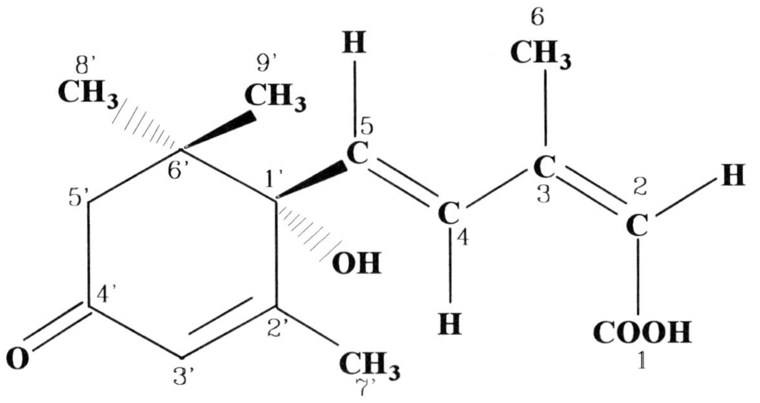

Figure 9.1 Structural formula of (+) abscisic acid.

9.2 Immunomodulation of abscisic acid activity

Abscisic acid (ABA) was recognized in the 1950s as a phytohormone affecting leaf abscission and bud dormancy. It was soon characterized as a sesquiterpene derived from mevalonate (Figure 9.1) although certain steps of its biosynthesis are still unknown. Continuing work using ABA-deficient and ABA-insensitive mutants has shown that it mediates many developmental and physiological processes, particularly during plant response to environmental stress, such as drought, high salinity or cold. ABA was found to be an essential mediator in triggering stomatal closure under conditions of water deficiency. Most of the available ABA-deficient mutants display an increased tendency to wilt (Tal and Nevo, 1973; Koornneef and Jorna, 1982; Quarrie, 1982; Wang *et al.*, 1984; Neill and Horgan, 1985; Parry *et al.*, 1991). Another quite different role of abscisic acid is regulation of development and maturation of seeds. This hormone is implicated in maintaining seed dormancy (Koornneef, 1986) and in the control of storage protein synthesis (Galau *et al.*, 1987; Koornneef *et al.*, 1989; Finkelstein and Somerville, 1990) and desiccation tolerance (Kermode and Bewly, 1987). Large numbers of ABA responsive genes have been identified in a variety of plant species (for reviews see Skriver and Mundy, 1990). Some of these genes are expressed during the normal embryogenic development programme when seeds desiccate and the embryos become dormant (Dure *et al.*, 1981). Under stress conditions, specific genes are expressed that can also be induced in non-stressed tissues by the application of exogenous ABA (Singh *et al.*, 1987; Gomez *et al.*, 1988; Mundy and Chua, 1988). The characterization of ABA-biosynthetic mutants however revealed that the ability of a given gene to respond to exogenous ABA under non-stress conditions does not necessarily imply that this gene is actually regulated by endogenous ABA upon stress (Gilmour and Thomashow, 1991; Nordin *et al.*, 1991). Results from studies involving application of exogenous ABA, or inhibitors of ABA synthesis, to plant cells are difficult to interpret, because it is very difficult to control uptake, translocation and metabolism of an applied substance.

A new approach presented here is based on blocking endogenous abscisic acid with specific scFv antibodies intracellularly expressed in transgenic plants. A prerequisite for this experiment was the availability of a suitable hybridoma cell line producing antibodies with high specificity and affinity to free abscisic acid (15-I-C5; Mertens *et al.*, 1983). The monoclonal antibody does not cross-react with (–)-*cis*, *trans*-ABA, (±)-*trans*, *trans*-ABA,

(±)-*cis*, *trans*-ABA methylester or with the main ABA precursors and metabolites such as xanthoxin, violaxanthin, phaseic and dihydrophaseic acids (Mertens, 1985). The scFv-encoding gene (scFv-ABA) was constructed from V_H and V_L cDNA, synthesized from mRNA isolated from the hybridoma cells, expressed in a bacterial expression system and characterized (Artsaenko *et al.*, 1994). Plant expression cassettes were designed to allow ubiquitous or seed-specific expression of the scFv gene. To study the pleiotropic physiological effects of abscisic acid in tobacco plants, the strong constitutive CaMV 35S promoter (Töpfer *et al.*, 1993) was used to regulate scFv-ABA expression (RA-cassette). Seed-specific expression cassettes containing the scFv-ABA gene under control of the *LeB4* (KD-cassette; Bäumlein *et al.*, 1986) and *USP* (BU-cassette; Bäumlein *et al.*, 1991) promoters were constructed to allow the investigation of seed-specific actions of the phytohormone. As a control for these experiments identical expression cassettes were constructed for the expression of an scFv antibody binding to the hapten oxazolone which is not found in plants (Fiedler and Conrad, 1995). To circumvent the problem of instability, the scFv antibodies were targeted to the lumen of the ER by fusion with the N-terminal legumin signal sequence (Saalbach *et al.*, 1991) and the C-terminal KDEL retention signal (Wandelt *et al.*, 1992). The expression cassettes were introduced into tobacco plants by *Agrobacterium*-mediated transformation and antibody accumulation levels and antigen-binding activity were investigated. In the RA transgenic plants, synthesis of the anti-ABA scFv antibody reached 6.8 per cent of total soluble protein (TSP) in leaves and 0.75 per cent in seeds. The KD and BU plants, respectively, accumulated up to 0.45 per cent and 4 per cent scFv of TSP exclusively in their seeds (Fiedler *et al.*, 1997). Plant scFv antibodies were shown in ELISA to bind free ABA with high affinity (Artsaenko *et al.*, 1995).

9.2.1 Ubiquitous expression of the anti-ABA scFv gene

Ubiquitous accumulation of functional anti-ABA scFv antibodies in transgenic tobacco plants caused a number of physiological effects which have been previously described for abscisic acid-deficient mutants. One of the most interesting phenomena observed was that transgenic plants wilted under ambient conditions (Figure 9.2). Many wilty mutants of higher plants have been isolated and characterized. Wilty, ABA-deficient mutants of tomato (*notabilis*, *flacca* and *sitiens*: Tal, 1966; Tal and Nevo, 1973; Neill and Horgan, 1985), potato (*droopy*: Quarrie, 1982), pea (*wilty*: Wang *et al.*, 1984), *Arabidopsis thaliana* (*aba*: Koornneef and Jorna, 1982; Karssen *et al.*, 1983), barley (*Az34*: Walker-Simmons *et al.*, 1989) and *Nicotiana plumbaginifolia* (*CKR1*: Blonstein *et al.*, 1991; Parry *et al.*, 1991; *aba1*: Leydecker *et al.*, 1995; *aba2*: Marin *et al.*, 1996) have been reported. Experiments with ABA-deficient mutants have elucidated the connection between the capacity to synthesize abscisic acid and the drought response of stomata. The mechanism of stomatal closure involves accumulation of abscisic acid in guard cells. Under the influence of an increased concentration of ABA, the cell turgor decreases and the stomata close. With the stomata closed, the rate of transpiration is reduced and the plant is able to conserve its internal water. The intermediate events of this process have been intensively studied, but models of ABA action are still incomplete, in part because ABA receptors have yet to be characterized (Schnabl and Ziegler, 1977; Schnabl, 1978; Hornberg and Weiler, 1984; Skriver and Mundy, 1990; Schroeder, 1992; Taylor *et al.*, 1995). The wilty phenotype of the RA transgenic plants expressing scFv-ABA was shown to be not due to ABA deficiency. Moreover, the transgenic plants had higher levels of abscisic acid in

(a)

(b)

(c)

Figure 9.2 Wilting of tobacco plants expressing the anti-ABA scFv antibodies in leaves. Transgenic and wild-type plants were grown in a greenhouse under plastic bags to maintain high humidity. Removal of the bags led to rapid wilting of the plants expressing scFv-ABA. RA 71, RA 60: transgenic plants expressing 2.4 per cent and 1 per cent anti-ABA scFv of total soluble protein, respectively; wild-type plant shown in the centre. (a) Immediately after removal of the bags; (b) after 30 minutes; and (c) after one hour at ambient humidity without bags.

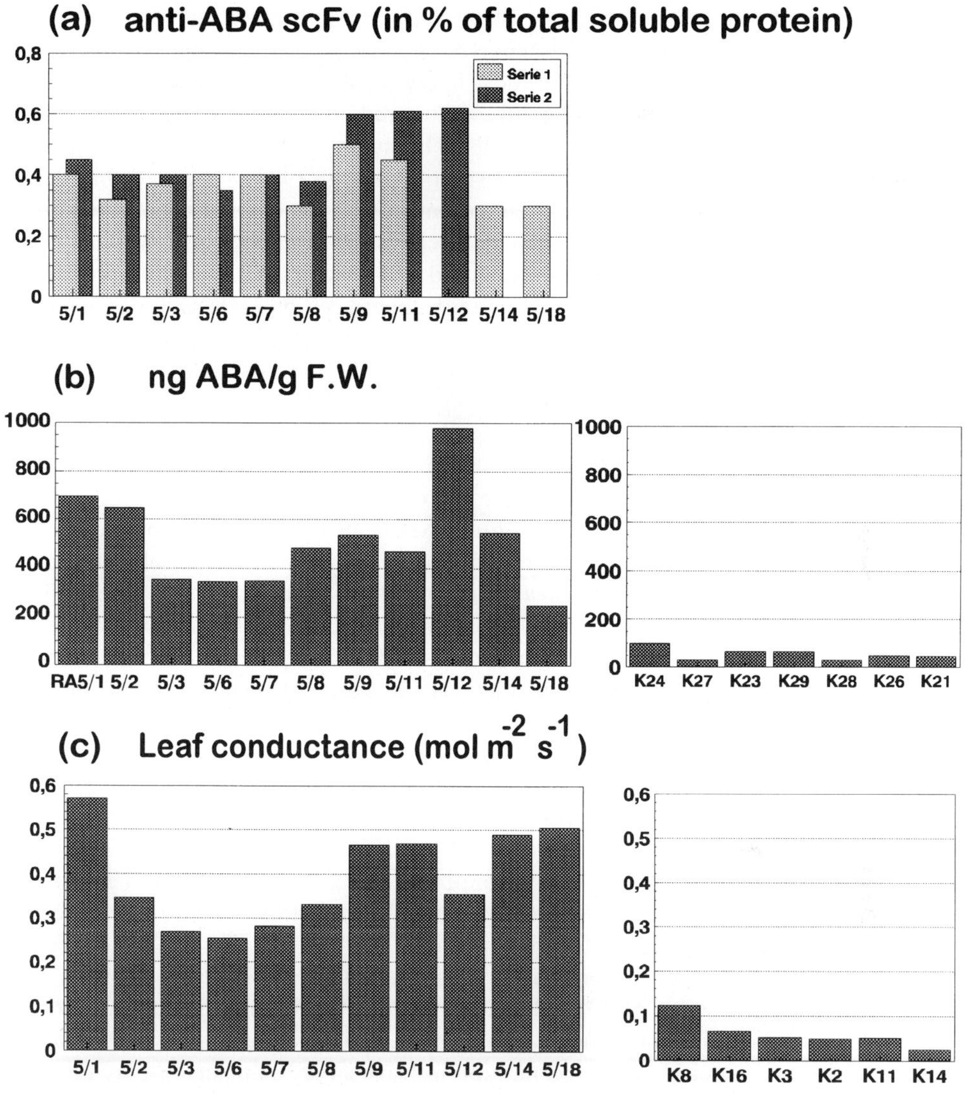

Figure 9.3 scFv antibody accumulation (a), abscisic acid content (b), and leaf conductance (c) measured in leaves of the T_1 transgenic plants – offspring of the RA5 primary transformant. Transgenic and wild-type plants were grown in a phytochamber at 95 per cent relative humidity. Determination of abscisic acid content and measurements of transpiration rates were performed as described by Artsaenko *et al.* (1995).

leaves than wild-type plants (Figure 9.3b). This could be explained by the disruption of the feedback regulation of ABA synthesis, or inaccessibility of ABA molecules to catabolize enzymes if they are sequestered in the ER away from the site of the enzyme action, or are physically protected by antibody binding. Studies are in progress to monitor intracellular distribution of abscisic acid in RA plants. A two- to threefold increase in ABA content in the walls of guard cells has been shown to be sufficient to induce closure of stomata (Hartung and Slovik, 1991).

In the RA transgenic plants the levels of abscisic acid in leaves were two- to tenfold higher than in wild-type plants. However, the leaf conductances of these plants were elevated compared with those of wild-type plants and not significantly altered by a decrease in CO_2 concentration (Figure 9.3c; Artsaenko *et al.*, 1995). This indicated that the plants were unable to close their stomata. The most likely mechanism to explain this phenomenon is that binding of abscisic acid by the scFv antibodies prevents its transport and interaction with ABA receptors in guard cells. The number of ABA receptors and their cellular location are at present an open question. Hartung (1983) suggested that it was the external, apoplastic ABA concentration in contact with the outer surface of the guard cell plasmalemma which influenced stomatal behaviour. Hornberg and Weiler (1984) demonstrated the presence of high-affinity, guard cell-specific, ABA-binding proteins at the outer surface of the plasmalemma of *Vicia faba* guard cell protoplasts. Later Schwartz *et al.* (1994) showed that ABA can act from within guard cells of *Commelina communis* to regulate both stomatal opening and stomatal closure. The results of Allan *et al.* (1994) also suggest a possible intracellular localization of the hypothesized ABA receptor. Although the subcellular location of ABA receptors is not known, the cytoplasmic perception of ABA molecules is discussed (see for review Allan and Trewavas, 1994). The high levels of anti-ABA scFv antibodies concentrated in the ER can provide an ABA sink in this compartment because the protonated form of ABA can easily permeate biomembranes (Hartung and Slovik, 1991). Accumulation of abscisic acid bound by the scFv antibodies in the endoplasmic reticulum may cause a deficiency of the phytohormone in other cellular compartments, including the cytoplasm. Thus, it is possible that ABA is prevented access to its site of action in the guard cells. Future studies addressing questions of the mechanism of ABA and anti-ABA scFv antibody interactions and their distribution in plant cells will include measurements of ABA and antibody concentrations in different cellular compartments, targeting of the scFv antibodies into the cytosol and the analysis of ABA accumulation after ABA treatment of transgenic and control plants. Moreover, the activity of ABA appears to depend on the presence of a free carboxyl group, a 2-*cis*, 4-*trans*-pentadienoic side chain, a 4′-ketone and a double bond in the cyclohexane ring (Walton, 1983).

Studies of the specificity of the 15-I-C5 antibody for the ABA analogues have shown that the structure of the side chain from carboxyl at C-1 to hydroxyl at C-1 is required for antibody binding (Figure 9.1; Walker-Simmons *et al.*, 1991). Therefore, ABA molecules bound by the anti-ABA scFv antibodies are most likely to be biologically inactive. This mimics the situation observed in ABA-deficient mutants: the RA transgenic plants are not able to close stomata and consequently are hypersensitive to water stress. Despite the increase of the total abscisic acid concentration in leaves, the RA plants display morphological characteristics typical for wilty ABA-deficient mutants, such as reduced apical dominance, adventitious roots and ‘waxy’ leaves (Artsaenko, 1996). The specific mode of the anti-ABA scFv antibody action in creating these changes was shown by comparing the RA plants with transgenic tobacco plants accumulating similar levels of anti-oxazolone scFv antibody in their leaves (Fiedler and Conrad, 1995). These plants did not wilt under normal greenhouse conditions and were morphologically indistinguishable from wild-type plants. This observation rules out the possibility that the expression of any scFv antibody is deleterious to the plants and confirms the specific action of the anti-ABA scFv antibody *in vivo*.

9.2.2 *Seed-specific expression of the anti-ABA scFv gene*

The application of transgene technology offers a unique possibility to modulate hormone levels and activity not only in the whole plant, but also in a tissue- or time-specific

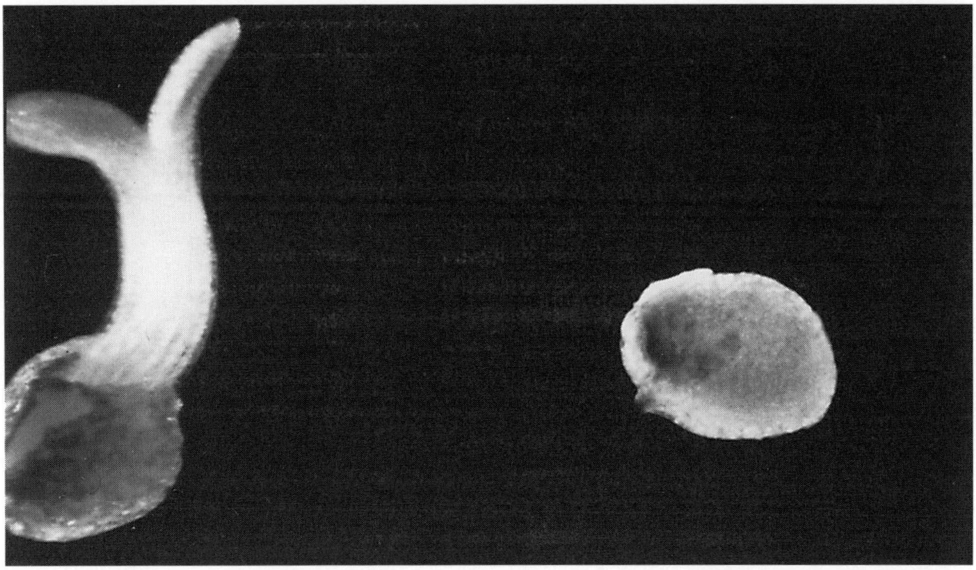

Figure 9.4 (a) Precocious germination of anti-ABA seed removed from its seed capsule 17 days after pollination and incubated for two days on filter paper soaked with water. (b) Similarly treated wild-type seeds did not germinate during this time.

manner. The ubiquitous expression of scFv-ABA throughout plant development caused a wide range of physiological changes leading to withering in ambient conditions of transgenic plants expressing high amounts of the scFv antibody. This affected in turn flowering, reproduction and seed formation. In the RA plants accumulating high amounts of the anti-ABA scFv antibody the quantity of seeds produced and their ability to germinate were reduced. In order to discriminate between the secondary effects caused by water vapour loss and the direct influence of ABA on seed development, transgenic plants expressing scFv-ABA exclusively in seeds were produced (Phillips *et al.*, 1997). These transgenic plants were not, apart from their seeds, phenotypically different from wild-type plants. The two promoters used for regulating the scFv expression in seeds (LeB4 and USP) show differences in strength and temporary activity during seed development as described by Fiedler *et al.* (1997). Under control of the USP promoter not only very high levels of antibody accumulation can be achieved, but it also allows synthesis of the scFv protein during earlier stages of embryo development compared with the expression driven by the LeB4 promoter. This period is shown to be critical and marks the beginning of the increase in ABA level in developing seeds. Although no apparent physiological effects were detected in the KD plants apart from an elevated ABA content in seeds, the development of BU transgenic seeds differed markedly from the normal embryo-development and seed-maturation programme. These transgenic seeds developed green cotyledons containing chloroplasts and accumulating photosynthetic pigments but produced lower levels of storage proteins and fewer oil bodies, while seeds expressing anti-oxazolone scFv had a normal phenotype and were indistinguishable from wild-type seeds (Phillips *et al.*, 1997). BU seeds germinated precociously if removed from seed capsules during development (Figure 9.4) but were not capable of germinating when allowed to develop until desiccation. Precocious germination, reduced desiccation tolerance and lower levels

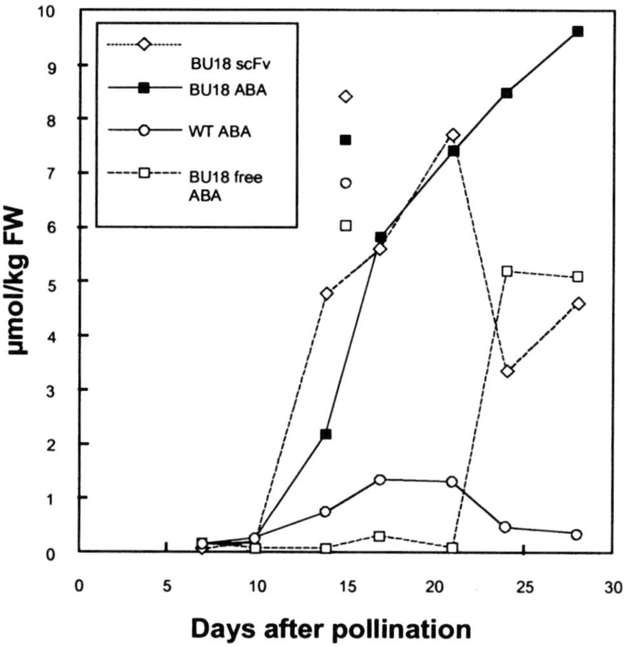

Figure 9.5 ABA levels in developing wild-type (WT ABA) and anti-ABA scFv seeds (BU18 ABA) compared with scFv levels in anti-ABA scFv seeds and free ABA calculated as described by Phillips *et al.* (1997). Data points are mean values for seeds from three plants of each type. The anti-ABA plants were offspring of primary transformants.

of seed storage protein accumulation were previously demonstrated for ABA-biosynthetic and insensitive mutants (Nambara *et al.*, 1992; Bäumlein *et al.*, 1994) although many of them are only mildly affected. The observation that binding of free ABA in BU seeds prevents accumulation of reserve proteins and causes reduction of desiccation tolerance constitutes new strong evidence for the important regulatory role of abscisic acid in controlling these processes, as underlined by Marion-Poll (1997). Another essential advantage of the immunomodulation of ABA activity is that these effects can be obtained without changes in biosynthetic pathways. In contrast to ABA-deficient mutants, the BU seeds accumulated total ABA to extremely high levels (up to 60-fold in mature seeds compared to wild type). The expected amount of free ABA in transgenic seeds was calculated based on the molar ratio of total ABA and scFv antibody and the apparent affinity constant of the anti-ABA scFv (Figure 9.5).

We could show that, despite the extremely high ABA levels, the amount of scFv antibody produced in transgenic seeds is sufficient to block ABA activity at the early to middle stages of embryo development, leading to a switch from the seed maturation programme to the germination programme.

9.3 Advantages and prerequisites of the phytohormone immunomodulation approach

Intracellular immunomodulation has important advantages over the use of existing mutants or conventional antisense technology to study hormone action. Antibody binding can inactivate an end-product of a biosynthetic pathway of a phytohormone without affecting the functions of any precursors. In contrast, in many biosynthetic mutants and during application of ABA inhibitors, ABA synthesis is blocked early in the pathway, thus making it difficult to attribute the observed physiological changes to a function of a particular endogenous regulator. The intracellular immunomodulation technology is applicable for any species which can be transformed and does not require cloning of genes involved in phytohormone synthesis. The levels of inhibition of phytohormone activity differ between individual transgenic plants due to variations in expression of the scFv gene depending on copy number, position effects and other factors. This may allow production of plants with different levels of biologically active phytohormone. Use of tissue- and development-specific promoters and intracellular targeting signals allows blocking of a phytohormone in specific organs, tissues, cells, and intracellular compartments at a given time during plant development. Furthermore, use of catalytic antibodies and bifunctional antibody fusion proteins could lead not only to a loss of function but also to engineering novel phenotypes.

A prerequisite for successful application of intracellular immunomodulation is the construction of a recombinant antibody highly specific for a target phytohormone and its thorough characterization. Any cross-reaction of the antibody with other endogenous plant substances may give misleading results. In addition, our data underline the importance of the choice of regulatory elements for the control of antibody synthesis and intracellular targeting.

9.4 Conclusions

Immunomodulation of the activity of a phytohormone as described in this chapter with the example of ABA demonstrates the great potential of intracellular immunomodulation technology for manipulation of plant physiology and development. Due to recent advances in phage display technology combined with improvements of plant transformation techniques and increasing choice of regulatory elements for time- and tissue-specific expression and intracellular targeting, the intracellular immunomodulation of plants will become a powerful tool in basic and applied plant research. We strongly believe that this technology will be widely used for both investigation of biological functions of regulatory molecules and creation of desired phenotypes with useful agronomic traits in the very near future.

Acknowledgements

The authors would like to thank I. Tillack, C. Liebing, E. Nagel and I. Heinemann for excellent technical assistance. The work described here was financially supported by the Bundesministerium für Bildung, Wissenschaft, Forschung und Technologie (BEO/210310261A) and by the Deutsche Forschungsgemeinschaft (Mu 925/1–2).

References

ALLAN, A.C. and TREWAVAS, A.J., 1994, Abscisic acid and gibberellin perception: inside or out?, *Plant Physiology*, **104**, 1107–1108.

ALLAN, A.C., FRICKER, M.D., WARD, J.L., BEALE, M.H. and TREWAVAS, A.J., 1994, Two transduction pathways mediate rapid effects of abscisic acid in *Commelina* guard cells, *The Plant Cell*, **6**, 1319–1328.

Artsaenko, O., 1996, Immunomodulation of ABA activity in transgenic tobacco plants, PhD Thesis, University of Halle.

Artsaenko, O., Weiler, E.W., Müntz, K. and Conrad, U., 1994, Construction and functional characterization of a single chain antibody binding to the plant hormone abscisic acid, *Journal of Plant Physiology*, **144**, 427–429.

Artsaenko, O., Peisker, M., zur Nieden, U., Fiedler, U., Weiler, E.W., Müntz, K. and Conrad., U., 1995, Expression of a single-chain Fv antibody against abscisic acid creates a wilty phenotype in transgenic tobacco, *The Plant Journal*, **8**, 745–750.

Bäumlein, H., Wobus, U., Pustell, J. and Kafatos, F.C., 1986, The legumin gene family: structure of a B type gene of *Vicia faba* and a possible legumin gene specific regulatory element, *Nucleic Acids Research*, **14**, 2707–2720.

Bäumlein, H., Boerjan, W., Nagy, I., Bassüner, R., Van Montagu, M., Inze., D. and Wobus, U., 1991, A novel seed protein gene from *Vicia faba* is developmentally regulated in transgenic tobacco and Arabidopsis plants, *Molecular General Genetics*, **225**, 459–467.

Bäumlein, H., Misera, S., Luerssen, H., Kolle, K., Horstmann, C., Wobus, U. and Müller, A.J., 1994, The *FUS3* gene of *Arabidopsis thaliana* is a regulator of gene expression during late embryogenesis, *The Plant Journal*, **6**, 379–387.

Biocca, S., Pierandrei-Amaldi, P. and Cattaneo, A., 1993, Intracellular expression of anti-p21ras single chain Fv fragments inhibits meiotic maturation of *Xenopus* oocytes, *Biochemical and Biophysical Research Communications*, **197**, 422–427.

Biocca, S., Ruberti, F., Tafani, M., Pierandrei-Amaldi, P. and Cattaneo, A., 1995, Redox state of single chain Fv fragments targeted to the endoplasmic reticulum, cytosol and mitochondria, *Bio/Technology*, **13**, 1110–1115.

Blonstein, A.D., Parry, A.D., Horgan, R. and King, P.J., 1991, A cytokinin-resistant mutant of *Nicotiana plumbaginifolia* is wilty, *Planta*, **183**, 244–250.

Duan, L., Bagasra, O., Laughlin, M.A., Oakes, J.W. and Pomerantz, R.J., 1994, Potent inhibition of human immunodeficiency virus type 1 replication by intracellular anti-Rev single-chain antibody, *Proceedings of the National Academy of Sciences USA*, **91**, 5075–5079.

Dure, L. III, Greenway, S.C. and Galau, G.A., 1981, Developmental biochemistry of cottonseed embryogenesis and germination. XIV. Changing mRNA populations as shown by *in vitro* and *in vivo* protein synthesis, *Biochemistry*, **20**, 4162–4168.

Fiedler, U. and Conrad, U., 1995, High-level production and long-term storage of engineered antibodies in transgenic tobacco seeds, *Bio/Technology*, **13**, 1090–1093.

Fiedler, U., Phillips, J., Artsaenko, O., Müntz, K. and Conrad, U., 1997, Optimization of scFv antibody production in transgenic plants, *Immunotechnology*, **3**, 205–216.

Finkelstein, R.R. and Somerville, C.R., 1990, Three classes of abscisic acid (ABA)-insensitive mutations of *Arabidopsis* define genes that control overlapping sub-sets of ABA responses, *Plant Physiology*, **94**, 1172–1179.

Galau, G.A., Bijaisoradat, N. and Hughes, D.W., 1987, Accumulation kinetics of cotton late embryogenesis-abundant mRNAs and storage protein mRNAs: coordinate regulation during embryogenesis and the role of abscisic acid, *Developmental Biology*, **123**, 198–212.

Gilmour, S.J. and Thomashow, M.F., 1991, Cold acclimation and cold-regulated gene expression in ABA mutants of *Arabidopsis thaliana*, *Plant Molecular Biology*, **17**, 1233–1240.

Gomez, J., Sanchez-Martinez, D., Stiefel, V., Rigau, J., Puigdomenech, P. and Pages, M., 1988, A gene induced by the plant hormone abscisic acid in response to water stress encodes a glycine-rich protein, *Nature*, **334**, 262–264.

Hartung, W., 1983, The site of abscisic acid in the guard cell plasmalemma of *Valerianella locusta*, *Plant, Cell and Environment*, **6**, 427–428.

Hartung, W. and Slovik, S., 1991, Physicochemical properties of plant growth regulators and plant tissues determine their distribution and redistribution: stomatal regulation by abscisic acid in leaves, *The New Phytologist*, **119**, 361–382.

Hornberg, C. and Weiler, E.W., 1984, High-affinity binding sites for abscisic acid on plasmalemma of *Vicia faba* guard cells, *Nature*, **310**, 321–325.

KARSSEN, C.M., DRINKHORST-VAN DER SWAN, D.L.C., BREEKLAND, A.E. and KOORNNEEF, M., 1983, Induction of dormancy during seed development by endogenous abscisic acid: studies on abscisic acid deficient genotypes of *Arabidopsis thaliana* (L.) Heynh., *Planta*, **157**, 158–165.

KERMODE, A.R. and BEWLEY, J.D., 1987, Regulatory processes involved in the switch from seed development to germination: possible roles for desiccation and ABA, *Drought Resistance in Plants, Physiological and Genetic Aspects*, MONTI, L. and PORCEDDU, E. (Eds), Brussels: EEC, pp. 59–76.

KOORNNEEF, M., 1986, Genetic aspects of abscisic acid, *Plant Genetic Research, A Genetic Approach to Plant Biochemistry*, Blonstein, A.D. and King, P.J. (Eds), Vienna: Springer-Verlag, pp. 35–54.

KOORNNEEF, M. and JORNA, M.L., 1982, The isolation of abscisic acid (ABA) deficient mutants by selection of induced revertants in non-germinating gibberellin sensitive lines of *Arabidopsis thaliana* (L.) Heynh., *Theoretical and Applied Genetics*, **61**, 385–393.

KOORNNEEF, M., HANHART, C.J., HILHORST, H.W.M. and KARSSEN, C.M., 1989, *In vivo* inhibition of seed development and reserve protein accumulation in recombinants of abscisic acid biosynthesis and responsiveness mutants in *Arabidopsis thaliana*, *Plant Physiology*, **90**, 463–469.

LEYDECKER, M.-T., MOUREAUX, T., KRAEPIEL, Y., SCHNORR, K. and CABOCHE, M., 1995, Molybdenum cofactor mutants, specifically impaired in xanthine dehydrogenase activity and abscisic acid biosynthesis, simultaneously overexpress nitrate reductase, *Plant Physiology*, **107**, 1427–1431.

MARASCO, W.A., HASELTINE, W.A. and CHEN, S., 1993, Design, intracellular expression and activity of a human anti-human immunodeficiency virus type 1 gp120 single-chain antibody, *Proceedings of the National Academy of Sciences USA*, **90**, 7889–7893.

MARIN, E., NUSSAUME, L., QUESADA, A., GONNEAU, M., SOTTA, B., HUGUENEY, P., *et al.*, 1996, Molecular identification of zeaxanthin epoxidase of *Nicotiana plumbaginiolia*, a gene in abscisic acid biosynthesis and corresponding to the ABA locus of *Arabidopsis thaliana*, *European Journal of Molecular Biology*, **15**, 2331–2342.

MARION-POLL, A., 1997, ABA and seed development, *Trends in Plant Sciences*, **2**, 447–448.

MERTENS, R., 1985, Herstellung und Einsatz monoklonaler Antikörper gegen Pflanzen-hormone, PhD Thesis, Ruhr-Universität Bochum.

MERTENS, R., DEUS-NEUMANN, B. and WEILER, E.W., 1983, Monoclonal antibodies for the detection and quantitation of the endogenous plant growth regulator, abscisic acid, *FEBS Letters*, **160**, 269–272.

MUNDY, J. and CHUA, N.-H., 1988, Abscisic acid and water-stress induce the expression of a novel rice gene, *The EMBO Journal*, **7**, 2279–2286.

NAMBARA, E., NAITO, S. and MCCOURT, P., 1992, A mutant of *Arabidopsis* which is defective in seed development and storage protein accumulation is a new *abi3* allele, *The Plant Journal*, **2**, 435–441.

NEILL, S.J. and HORGAN, R., 1985, Abscisic acid production and water relations in wilty tomato mutants subjected to water deficiency, *The Journal of Experimental Botany*, **36**, 1222–1231.

NORDIN, K., HEINO, P. and PALVA, E.T., 1991, Separate signal pathways regulate the expression of a low-temperature induced gene in *Arabidopsis thaliana* (L.) Heynh., *Plant Molecular Biology*, **16**, 1061–1071.

OWEN, M.R., GANDECHA, A., COCKBURN, W. and WHITELAM, G.C., 1992, Synthesis of a functional anti-phytochrome single-chain Fv protein in transgenic tobacco, *Bio/Technology*, **10**, 790–794.

PARRY, A.D., BLONSTEIN, A.D., BABIANO, M.J., KING, P.J. and HORGAN, R., 1991, Abscisic acid metabolism in a wilty mutant of *Nicotiana plumbaginifolia*, *Planta*, **183**, 790–794.

PHILLIPS, J., ARTSAENKO, O., FIEDLER, U., HORSTMANN, C., MOCK, H.-P., MÜNTZ, K. and CONRAD, U., 1997, Seed-specific immunomodulation of abscisic acid activity induces a developmental switch, *European Journal of Molecular Biology*, **16**, 4489–4496.

QUARRIE, S., 1982, DROOPY: a wilty mutant of potato deficient in abscisic acid, *Plant Cell Environment*, **5**, 23–26.

RICHARDSON, J.H., SODROSKI, J.G., WALDMANN, T.A. and MARASCO, W.A., 1995, Phenotypic knockout of the high-affinity human interleukin 2 receptor by intracellular single-chain antibodies against the a subunit of the receptor, *Proceedings of the National Academy of Sciences USA*, **92**, 3137–3141.

SAALBACH, G., JUNG, R., KUNZE, G., SAALBACH, I., ADLER, K. and MÜNTZ, K., 1991, Different legumin protein domains act as vacuolar targeting signals, *The Plant Cell*, **3**, 695–708.

SCHNABL, H., 1978, The effect of C1- upon the sensitivity of starch-containing and starch-deficient stomata and guard cell protoplasts towards potassium ions, fusicoccin and abscisic acid, *Planta*, **144**, 95–100.

SCHNABL, H., and ZIEGLER, H., 1977, The mechanism of stomatal movement in *Allium cepa* L., *Planta*, **136**, 37–43.

SCHROEDER, J.I., 1992, Plasma membrane ion channel regulation during abscisic acid-induced closing of stomata, *Philosophical Transactions of the Royal Society*, **338**, 83–89.

SCHWARTZ, A., WU, W.-H., TUCKER, E.B. and ASSMANN, S.M., 1994, Inhibition of inward K^+ channels and stomatal response by abscisic acid: an intracellular locus of phytohormone action, *Proceedings of the National Academy of Sciences USA*, **91**, 4019–4023.

SINGH, N.K., LA ROSA, P.C., HANDRA, A.K., HASEGAWA, P.M. and BRESSAN, R.A., 1987, Hormonal regulation of protein synthesis associated with salt tolerance in plant cells, *Proceedings of the National Academy of Sciences USA*, **84**, 739–743.

SKRIVER, K. and MUNDY, J., 1990, Gene expression in response to abscisic acid and osmotic stress, *The Plant Cell*, **2**, 503–512.

TAL, M., 1966, Abnormal stomatal behaviour in wilty mutants of tomato, *Plant Physiology*, **41**, 1387–1391.

TAL, M. and NEVO, Y., 1973, Abnormal stomatal behaviour and root resistance, and hormonal imbalance in three wilty mutants of tomato, *Biochemical Genetics*, **8**, 291–300.

TAVLADORAKI, P., BENVENUTO, E., TRINCA, S., MARTINIS, D., CATTANEO, A. and GALEFFI, P., 1993, Transgenic plants expressing a functional single-chain Fv antibody are specifically protected from virus attack, *Nature*, **36**, 469–472.

TAYLOR, J.E., RENWICK, K.F., WEBB, A.A., MCAINSH, M.R., FURINI, A., BARTELS, D., *et al.*, 1995, ABA-regulated promoter activity in stomatal guard cells, *The Plant Journal*, **7**, 129–134.

TÖPFER, R., MAAS, C., HÖRICKE-GRANDPIERRE, C., SCHELL, J. and STEINBISS, H.-H., 1993, Expression vectors for high-level expression in dicotyledonous and monocotyledonous plants, *Methods in Enzymology*, **217**, 66–78.

VOSS, A., NIERSBACH, M., HAIN, R., HIRSCH, H.J., LIAO, Y., KREUZALER, F. and FISCHER, R., 1995, Reduced virus infectivity in *N. tabacum* secreting a TMV-specific full size antibody, *Molecular Breeding*, **1**, 39–50.

WALKER-SIMMONS, M.K., KUDRNA, D.A. and WARNER, R.I., 1989, Reduced accumulation of ABA during water stress in a molybdenum cofactor mutant of barley, *Plant Physiology*, **90**, 728–733.

WALKER-SIMMONS, M.K., REANEY, M.J., QUARRIE, S.A., PERATA, P., VERNIERY, P. and ABRAMS, S.R., 1991, Monoclonal antibody recognition of abscisic acid analogs, *Plant Physiology*, **95**, 46–51.

WALTON, D.C., 1983, Structure–activity relationships of abscisic acid analogs and metabolites, in *Abscisic Acid*, Addicot, F.T. (Ed.), Praeger, New York, pp. 113–146.

WANDELT, C.I., KHAN, M.R.I., CRAIG, S., SCHROEDER, H.E., SPENCER, D. and HIGGINS, T.J.V., 1992, Vicilin with carboxyterminal KDEL is retained in the endoplasmic reticulum and accumulates to high levels in the leaves of transgenic plants, *The Plant Journal*, **2**, 181–192.

WANG, T.L., DONKIN, M.E. and MARTIN, E.S., 1984, The physiology of a wilty pea: abscisic acid production and water stress, *The Journal of Experimental Botany*, **35**, 1222–1232.

10

Engineering of *Beet necrotic yellow vein virus* (BNYVV) Resistance in *Nicotiana benthamiana*

LOTHAR F. FECKER AND RENATE KOENIG

10.1 Introduction

The natural ability of animals to produce antibodies against many pathogens, including viruses, often protects the animal from reinfection by the pathogen. Unfortunately the initial infection has often already exerted its deleterious effects before the protective antibodies have been produced. Man has learned to artificially stimulate antibody-mediated protection with the use of vaccines consisting of attenuated or killed forms of a pathogen ('active immunization'). The advent of genetic engineering has made it possible to produce antibodies in plants, independently of any natural immune system. Plants may be used to produce antibodies or antibody-derived molecules such as single-chain fragments (scFv) in which the variable domains of antibody heavy and light chains (V_H and V_L) are joined together by a linker peptide to form a single molecule. The various types of 'plantibodies' have many potential uses. They may serve as commercially valuable reagents in which case the plants are merely a 'biofactory' for the production of large quantities of the reagent, from which the reagent is then harvested ('antibody farming'). Alternatively, the antibodies may be expressed in plants for purposes of 'immunomodulation'. Plant-expressed antibodies may be designed to modify or block the activity, function or mobility of the cognate antigens (Whitelam *et al.*, 1994). If the antigen is a pathogen or a pathogen-associated non-structural protein, its blockage may render the plant resistant to infection. This antibody-mediated resistance is an attractive alternative to the various forms of pathogen-derived resistance, for which unintended side effects such as heteroencapsidations and recombinations of viral genomes cannot always be excluded (Farinelli *et al.*, 1992; Tepfer, 1993; Greene and Allison, 1994; Wintermantel and Schoelz, 1996).

Choosing the right antibody for inducing resistance may be a difficult task. Many of the non-structural proteins of a virus are needed in much lower concentrations than the structural ones. It would therefore seem easier to interfere with virus multiplication by blocking a non-structural protein than a structural one. However, for most non-structural proteins little is known about their natural conformation(s) in the plant cell. They may occur as 'free' proteins, for example, in *statu nascendi* or they may interact either with other proteins (for example, in membranes or enzyme complexes), or with nucleic acids. Antibodies raised against proteins expressed in bacteria may be specific only for epitopes of the denatured protein and may not recognize the protein in its native state within the plant cell.

For *Tobacco mosaic virus* (TMV) (genus *Tobamovirus*) it has been shown that its cotranslational disassembly is much more efficiently inhibited by antibodies to some metatopes (epitopes present on both dissociated and polymerized forms of the viral coat protein), than by antibodies to some neotopes (epitopes specific for the quarternary structure of the TMV particles) (Saunal *et al.*, 1993). Antibodies which bind to those metatopes located closest to the central axis in the polymerized particles and which are exposed on one particle extremity are inhibitory (Saunal and van Regenmortel, 1995). This is discussed in Chapter 1.

When considering antibody-mediated resistance, it should also be kept in mind that antibodies or antibody-derived molecules which are expected to block the multiplication of RNA viruses should be targeted to the cytoplasm, because this is the compartment where these viruses multiply. This requirement, however, is not always easy to fulfil (see following sections of this chapter).

Antibody-mediated resistance has previously been achieved for TMV and for *Artichoke mottled crinkle virus* (AMCV) (genus *Tombusvirus*). Voss *et al.* (1995) observed that transgenic tobacco plants expressing TMV coat protein-specific antibodies developed fewer local lesions than the non-transgenic control plants after inoculation with TMV. For correct folding and assembly the heavy and light chains of the antibody were targeted by the immunoglobulin signal peptide to the endoplasmic reticulum (ER). From this compartment the antibodies were transported to the intercellular fluid where they accumulated. Tavladoraki *et al.* (1993) have expressed AMCV coat protein-specific scFv (without a targeting signal), in the cytoplasm of transgenic *N. benthamiana*. These plants became partially resistant to infections by AMCV.

We have studied the potential of scFv specific for *Beet necrotic yellow vein virus* (BNYVV) (genus *Benyvirus*) to interfere with the multiplication and pathogenicity of this virus. BNYVV is the causal agent of sugar beet rhizomania, one of the economically most important sugar beet diseases. It is transmitted by the soil-borne fungus *Polymyxa betae.* Its genome consists of four to five single-stranded positive-sense RNA species (Richards and Tamada, 1992; Koenig and Lesemann, 1999). Hybridoma cell lines producing Mabs specific either for the BNYVV 25 kDa non-structural protein (25K protein) or for epitope 4b (located close to the C-terminus of the BNYVV coat protein and exposed along the entire length of the particles) (Commandeur *et al.*, 1994) served as sources for obtaining scFv-encoding sequences (Kaufmann, 1995). The 25K protein is encoded on BNYVV RNA 3, which is mainly responsible for the beard formation and the yield losses observed in sugar beet after infection with BNYVV (Tamada *et al.*, 1990; Koenig *et al.*, 1992). *N. benthamiana* was chosen as a model plant because it is more readily transformed than sugar beet and it can be infected by some isolates of BNYVV (Fecker *et al.*, 1997).

10.2 Cloning strategy and expression of different BNYVV-specific scFv in *Escherichia coli*

Messenger RNA was isolated from four hybridoma cell lines and RT-PCR was used to amplify the coding sequences for the respective variable regions of the heavy and light chains. The PCR products were cloned separately into the bacterial expression vector pCOCK (Engelhardt *et al.,* 1994) as described in Chapters 2 and 4. Our pCOCK-derived

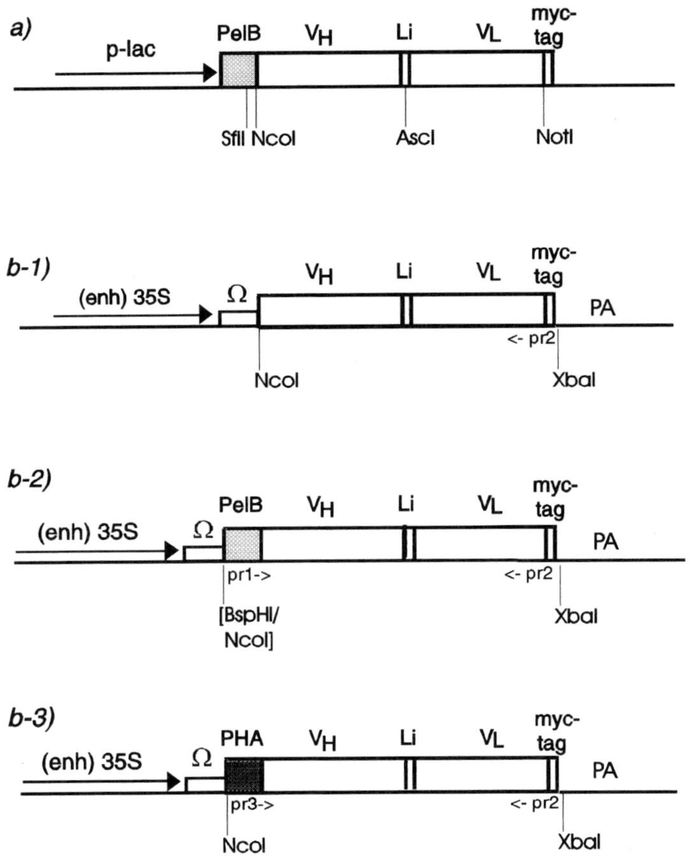

Figure 10.1 Gene constructs for the expression of the BNYVV coat protein- and 25K protein-specific scFv in *E. coli* (a) and in plants (b-1 to b-3). The bacterial scFv expression cassettes were established in the vector pCOCK (Engelhardt *et al.*, 1994). For plant transformation the scFv-encoding sequences were integrated into the cauliflower mosaic virus 35S promoter expression cassettes of either pRT103 (35S promoter) (Töpfer *et al.*, 1987) or pRT213 (enhanced 35S promoter and omega leader region of tobacco mosaic virus), respectively. (b-1) Construct without the coding sequence for a signal peptide for the expression of the scFv in the cytoplasm. (b-2 and b-3) Constructs containing the coding sequences for the PelB or PHA signal peptides, respectively, for targeting the scFv to the secretory pathway. Abbreviations: p-lac, lac promoter; (enh)35S, (enhanced) 35S promoter; , omega leader region of tobacco mosaic virus; PA, polyadenylation signal; PelB, PHA, V_H, V_L, Li, myc-tag: DNA sequences encoding the PelB or PHA signal peptides, the variable regions of the antibody heavy and light chains, the linker and the myc-tag peptides, respectively. PCR primers: pr1 = PelBBackBspHI; pr2 = VKForXbaI; pr3 = PHABackNcoI. The restriction sites necessary for cloning are indicated.

clones (Figure 10.1a) provided a means of sequencing the scFv genes, and of characterizing the properties of the proteins prior to transforming plants.

Functionally active scFv were obtained only from one cell line for each of the two proteins (Figure 10.2a). When expressed in HB2151 *E. coli* cells, the BNYVV coat protein-specific scFv was found to be present predominantly in the bacterial culture fluid whereas the 25K protein-specific scFv was mainly retained in the periplasm. Considerable

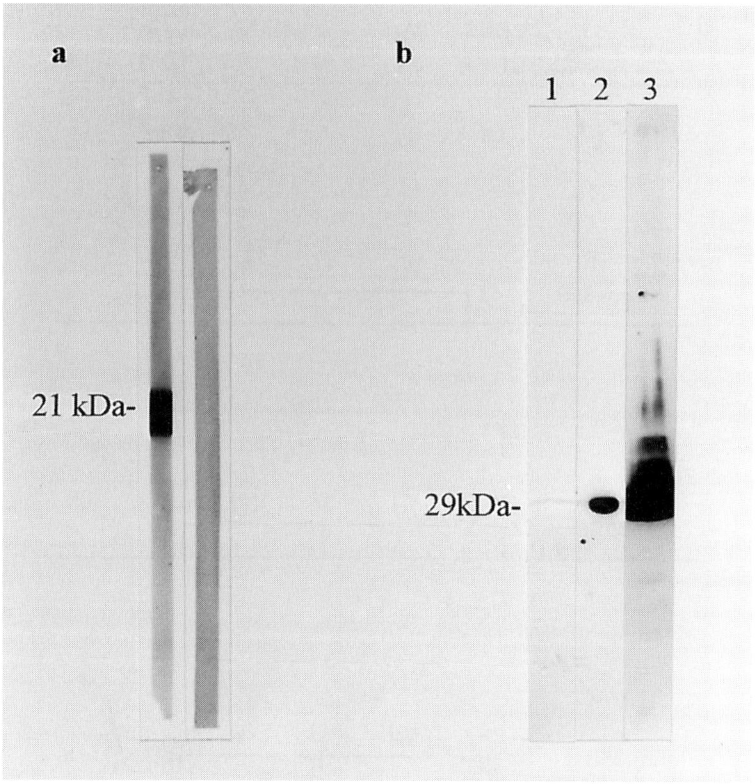

Figure 10.2 (a) Reactivity of BNYVV 21 kDa coat protein immobilized on nitrocellulose strips by Western blotting with its cognate scFv in culture fluids of *E. coli* cultures (left). No reactivity was seen with the culture fluid from a non-expressing *E. coli* control culture (right). The binding of the scFv was detected by means of the myc-tag-specific Mab 9E10 and rabbit anti-mouse IgG alkaline phosphatase-conjugated antibodies. (b) Attempts to detect the 29 kDa scFv in the culture fluid (1), in the cell lysate (2), and in the insoluble membrane fraction (3) of an *E. coli* culture after Western blotting. Only the scFv in soluble fractions are able to bind to BNYVV coat protein. Detection as in (a).

amounts of both scFv were also found in the insoluble cell fractions (shown for the coat protein-specific scFv in Figure 10.2b).

The DNA coding sequences for the coat protein- and the 25K protein-specific scFv were determined (EMBL Database accession numbers Z70661 and Z70662, respectively). The coat protein-specific V_H and V_L regions are in subgroups III(C) and I, and the 25K protein-specific V_H and V_L regions belong to groups II(B) and III (Kabat *et al.*, 1991).

In dot blot immunoassays the two scFv allowed the detection of their immobilized homologous antigens with almost the same sensitivity as the original Mab from which they were derived (0.2–0.4 ng coat protein and 0.8 ng 25K protein). There was no cross-reactivity with non-infected plant extracts or other antigens in this type of test. In ELISA on plates precoated with polyclonal antibodies, the coat protein-specific scFv also allowed the highly sensitive and specific detection of BNYVV particles; the 25K protein-specific scFv, on the other hand, showed some additional reactivity with extracts from healthy plants.

10.3 Transformation of *N. benthamiana* with various scFv-encoding constructs

Three types of gene constructs were designed for each of the two scFv (Figure 10.1b-1 to b-3). The respective pCOCK-derived cDNA clones (Figure 10.1a) served as starting material for amplifying the coding sequences. For targeting the scFv to the secretory pathway and to ensure correct folding, either the bacterial pelB signal peptide (Lei *et al.*, 1987) (Figure 10.1b-2) or the *Phaseolus vulgaris* phytohaemagglutinin signal peptide (PHA) (Dorel *et al.*, 1989; Hunt and Chrispeels, 1991) (Figure 10.1b-3) was included. For cytoplasmic expression, the sequence for the signal peptide (Figure 10.1b-1) was omitted. The bacterial pelB signal peptide has previously been shown to have some activity also in higher plants (Benvenuto *et al.*, 1991). The DNA fragments (Figure 10.1b-1 to b-3) were integrated either between the CaMV 35S promoter and the polyadenylation site of pRT103 (Töpfer *et al.*, 1987) or between the enhanced 35S promoter (Ow *et al.*, 1987) connected to the TMV Ω-leader region (Gallie *et al.*, 1987) and the CaMV polyadenylation signal of pRT213 (Figure 10.1b-1 to b-3) resulting in expression cassettes from the CaMV 35S promoter to the CaMV polyadenlylation signal (Figure 10.1b-1 to b-3). These were integrated into the binary vector pLX222 (Landsmann *et al.*, 1988) for the transformation of *Agrobacterium tumefaciens* and *N. benthamiana.*

10.3.1 Procedures

The coding sequences for the BNYVV coat protein- or 25K protein-specific scFv together with the coding sequences for the pelB signal peptide and the myc-tag were amplified from the respective pCOCK-derived plasmid clones (Figure 10.1a) by PCR using the primers pelBBackBspHI (5′-dGACAGTCATCATGAAATACC-3′) and VKForXbaI (5′-dTATGCTCTAGATTCAACAGTCTATGCGGC-3′). Aliquots of the amplified DNA were digested with *Nco* I to remove the signal peptide-encoding sequence, or with *Bsp*H I to produce DNA fragments which still contained the pelB signal peptide-encoding sequence. A third type of DNA fragment was obtained by amplifying the scFv-encoding sequences from the respective pCOCK-derived plasmids (Figure 10.1a) using the 100-mer primer PHABackNcoI (5′-dATCAGCCATGGCTTCCTCCAAGTTC-TTCACTGTCCTCTTCCTTGTGCTTCTCACCCACGCAAACTCAAGCAACGATSAG-GTCCAGCTGCAGSAGTCWGG-3′; S = C or G; M = A or C; R = A or G; W = A or T) together with the primer VKForXbaI. The amplified DNA was first digested with *Nco I.* All three types of DNA fragments were then digested with *Xba* I and integrated between the *Nco* I and *Xba* I restriction sites of either pRT103 or pRT213. The expression cassettes from the pRT vectors were isolated and integrated into the binary vector pLX222. *A. tumefaciens* LBA4404 Rif^r (Hoekema *et al.*, 1983) was transformed as described by An *et al.* (1988) and used for transforming *N. benthamiana* by a modified leaf disc method (Horsch *et al.*, 1985). Extracts from the regenerated kanamycin-tolerant plants were tested for neomycin phosphotransferase by ELISA (5Prime > 3Prime, Inc.) and their protein content was determined by means of the BCA protein assay reagent (Pierce). The amounts of antigen-binding scFv in plant extracts were estimated by immunodots using antigen-coated nylon membrane discs (Qiagen). Detection of the scFv binding was by means of the Mab 9E10 and rabbit anti-mouse IgG alkaline phosphatase-conjugated antibodies. The colour intensities obtained after the addition of substrate were compared with those obtained from antigen-coated discs exposed to bacterially-derived solutions containing known amounts of the denatured scFv and detected in the same way.

Table 10.1 Detection of the expression of BNYVV coat protein- and P25-specific scFv in *Nicotiana benthamiana* transformed with various gene constructs

Specificity of scFv	pRT vector	Signal peptide in the expression cassette	Leaf discs treated with transformed Agrobacteria	Kanamycin resistant plants – regenerated	Plants with		
					strong[a] scFv production	medium[a] or low[a] scFv production	no detectable scFv production[b]
BNYVV cp	103	None	60	39	None	None	39
	213	None	60	10	None	None	10
	103	PelB	60	5	None	3	2
	213	PelB	60	3	None	None	3
	103	PHA	60	25	None	7	18
	213	PHA	60	13	4	7	2
BNYVV P25	103	None	30	None	None	None	None
	213	None	30	13	None	None	13
	103	PelB	60	None	None	None	None
	213	PelB	60	None	None	None	None
	103	PHA	30	21	1	None	20
	213	PHA	30	8	2	None	6

[a] The scFv concentrations were approximately 0.002 per cent to 0.01 per cent in plants with 'strong scFv production' and <0.002 per cent of total protein in the plants with 'medium or low scFv production'.
[b] Positive results in the neomycin phosphotransferase test indicated that these plants were transformed.

10.3.2 Results – targeting scFv

With *N. benthamiana*, scFv production was recorded mainly in plants transformed with constructs containing the coding sequence for the PHA signal peptide (Table 10.1). The scFv expression levels varied between 0.001 per cent and 0.01 per cent of the total soluble protein (TSP). Low levels of expression (0.001 per cent or less of the TSP) of BNYVV coat protein-specific scFv were recorded in three plants which had been transformed with the construct in which the scFv-encoding sequence was preceded by the coding sequence for the pelB signal peptide (Figure 10.1b-2; Table 10.1). Sixty-two kanamycin-resistant plants transformed with constructs for the cytoplasmic expression of scFv specific either for BNYVV 25k or coat protein were produced (Figure 10.1b-1), but no scFv were detected.

Concentrations of the coat protein-specific scFv in four plants were equal to those in the *E. coli* culture fluids described in the previous section. As little as 0.2 to 0.4 ng of BNYVV coat protein was detected by a four-fold diluted extract obtained from leaf tissue of an scFv-expressing plant. After selfing those plants with the highest scFv levels, plants of the F_2 or F_3 generation, were selected with an scFv content of 0.1 per cent of the TSP.

Cell suspension cultures were established from transformed plants expressing coat protein-specific scFv. No scFv were detected in the culture fluids 48 hours after transferring the suspension cultures to fresh medium (Figure 10.3). However, functionally active scFv were detected in the supernatants of cell homogenates of suspension cultures derived

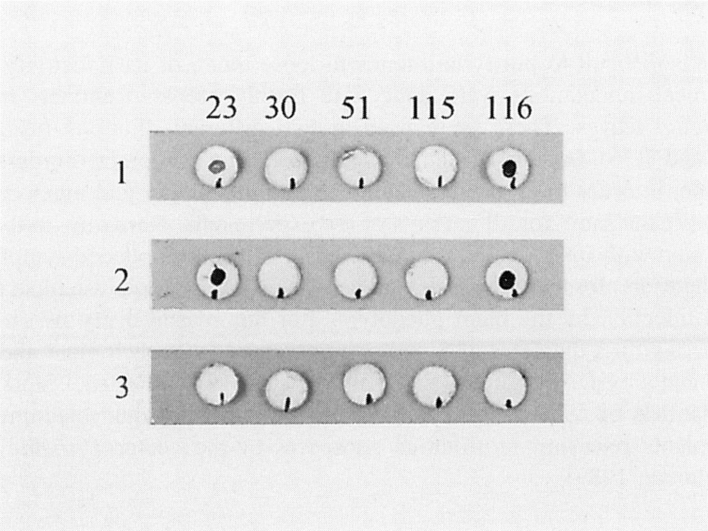

Figure 10.3 Attempts to detect BNYVV coat protein-specific scFv in cellular fractions (row 1), cellular fractions without cell walls (row 2), and in culture fluids (row 3) of cell suspension cultures derived from the transgenic *N. benthamiana* lines 23, 30, 51, 115, and 116. Plants of lines 23 and 116 were transformed with constructs containing the coding sequences for the PHA signal peptide, plants of lines 30 and 115 with constructs containing the coding sequences for the pelB signal peptide, and plants of line 51 with constructs lacking a signal peptide encoding sequence, resulting in cytoplasmic expression. The scFv were bound to immobilized BNYVV coat protein on the filter discs and detected by means of the myc-tag-specific Mab 9E10 and rabbit anti-mouse IgG alkaline phosphatase-conjugated antibodies.

from scFv-expressing plants in which the PHA signal peptide had been used for targeting. These results indicate that the coat protein-specific scFv which were expressed with the PHA signal peptide were retained in the cells, presumably in the ER or an ER-associated compartment rather than being secreted into the extracellular space.

10.4 Influence of the expression of BNYVV coat protein-specific scFv in transgenic *N. benthamiana* on the early and late stages of infection with BNYVV

To find whether the expression of BNYVV coat protein-specific scFv in *N. benthamiana* plants offers a protection against infection by BNYVV and/or against its pathogenic effects, we tested the progeny of two transgenic plants which after selfing had produced seeds. These plants of the F_1 or F_2 generation showed higher levels of scFv expression than the original plants of the F_0 generation. The amount of scFv in the TSP was 0.005 to 0.01 per cent for the progeny of line 23 and 0.05 to 0.1 per cent for that of line 116. The growth rate was normal for the plants of line 23, but it was somewhat reduced for the strongly expressing line 116.

The plants which expressed BNYVV 25K protein-specific scFv were not inoculated with BNYVV, because they grew only poorly without forming roots.

10.4.1 *Procedures*

As BNYVV is difficult to purify and tends to loose much of its infectivity during storage, mechanical inoculations were done with freshly prepared extracts from infected *N. benthamiana* leaves. These were used at two different dilutions (usually 1:4 and 1:16 or 1:6 and 1:36). Inoculated plants were placed in a randomized order on a growth chamber table, in order to ensure that the average light supply and other environmental conditions were the same for all variants of the experiments. Normally, in the first weeks after inoculation with BNYVV, *N. benthamiana* shows no pronounced symptoms of virus infection. Therefore, for the transgenic plants it was not possible to establish the influence of the virus infection by the plant phenotype. For this reason every two to three days, three leaf discs with a diameter of 5 mm were obtained from each plant and were tested for virus contents by DAS ELISA (Clark and Adams, 1977). For root inoculations with BNYVV, plantlets of *N. benthamiana* were co-cultivated in liquid medium with young sugar beet plants releasing viruliferous zoospores of the vector *P. betae* (Peters and Godfrey-Veltman, 1989).

10.4.2 *Results – influence of scFv on the multiplication of BNYVV and its pathogenic effects*

The influence of BNYVV infections on plants expressing BNYVV coat protein-specific scFv was studied with two lines of transformed plants, i.e. 23 and 116, in a total of eight experiments done at different times during the year.

In the early stages, after mechanical inoculation, the percentage of plants with detectable amounts of BNYVV was usually lower for the scFv-expressing lines than for non-transgenic plants or transgenic plants in which we had failed to detect scFv formation

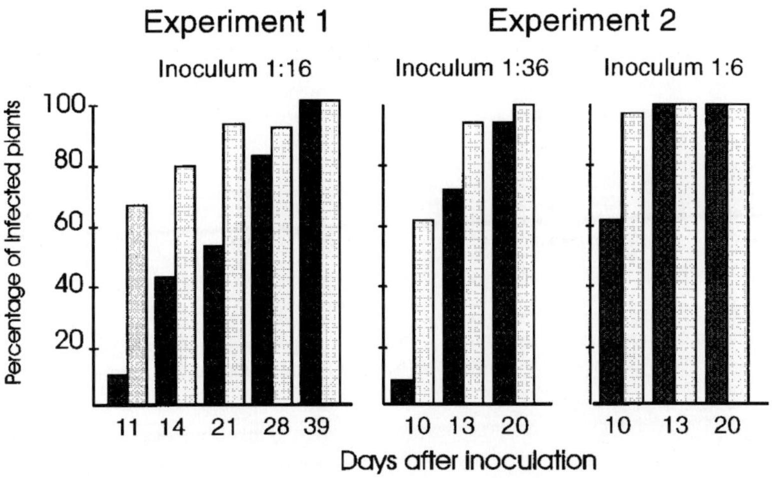

Figure 10.4 Percentage of *N. benthamiana* plants in which BNYVV infections could be detected by ELISA at various times after mechanical inoculation. Shaded boxes, plants of the transgenic line 23 in which BNYVV coat protein-specific scFv are expressed in the ER or an ER-associated compartment; open boxes, non-expressing control plants. Experiment 1 was performed with ten transgenic and ten control plants inoculated with 1:16 diluted crude sap from infected *N. benthamiana*. In Experiment 2, 18 transgenic and 18 control plants were inoculated with 1:36 diluted sap and the same number of transgenic and control plants were inoculated with 1:6 diluted sap.

(Table 10.1). The length of time in which the percentage of infection differed between the scFv-expressing plants and the controls varied somewhat between experiments. In some experiments (for example, Figure 10.4, Experiment 1) a lower percentage of infection was detectable for at least three weeks after inoculation; in others, however, clear differences in the infection rates were observed for only about two weeks or even less (Figure 10.4, Experiment 2).

Differences in the percentages of infection were always much more pronounced when a dilute virus inoculum was used rather than a more concentrated one (Figure 10.4, Experiment 2). In all experiments, all inoculated scFv-expressing plants did eventually become infected. With line 23 the average virus concentration reached in the transgenic plants equalled that observed in the control plants, but with line 116 it tended to be somewhat lower.

During the first weeks after inoculation we observed no pronounced differences between the phenotypes of the scFv-expressing plants of line 23 and the control plants. In the later stages (eight to ten weeks after inoculation) the scFv-expressing plants tended to look more vigorous than the non-expressing control plants (Figure 10.5). There was no strict correlation between the average length of time during which infection was delayed in the scFv-expressing plants and the amount of protection offered by the scFv against virus-induced symptoms in the late infection stages. For instance, in the plants depicted on the right side of Figure 10.5 in which this protective effect had been rather pronounced, a delay of infection was observed only up to the second week after inoculation.

In one experiment we have also studied the transmission of BNYVV to *N. benthamiana* via the roots by the natural vector *P. betae*. Seventeen days after inoculation only two of 16 plants of the transgenic line 23, but six of 14 non-transgenic plants were identified as

Figure 10.5 Phenotype of *N. benthamiana* plants of the transgenic line 23 (left side) in which BNYVV coat protein-specific scFv are expressed in the ER or an ER-associated compartment and of non-transgenic control plants (right side) ten weeks after mechanical inoculation with 1:36 diluted crude sap from BNYVV-infected *N. benthamiana.*

being BNYVV-infected in the roots. After 30 days the virus was detected in all transgenic and non-transgenic plants, and no significant differences in the average virus concentration were observed between the two groups. Thus, also after inoculation with *P. betae* there was a delay of BNYVV infection in the transgenic plants.

10.5 Discussion

As discussed earlier in this chapter, to block the multiplication of an RNA virus, its cognate antibodies or scFv should be expressed in the cytoplasm of the plant cells. To achieve this we transformed plants with expression cassettes containing the scFv-encoding sequences without the coding sequence for a signal peptide (Figure 10.1b-1). This strategy had proved successful for the AMCV coat protein-specific scFv (Tavladoraki *et al.*, 1993). However, for scFv specific for either the coat protein or the non-structural 25K protein of BNYVV we were unable to detect any cytoplasmic scFv expression in the plants. We cannot exclude the possibility, though, that some of our plants produced denatured scFv which, in contrast to the AMCV-specific scFv, were highly sensitive to proteolytic digestion in the reducing environment of the cytoplasm. This environment (Biocca *et al.*, 1995) may have impaired the formation of the disulphide bonds necessary for the correct folding of the scFv. Reasonable concentrations (0.002 to 0.01 per cent of the TSP) of BNYVV coat protein or 25K protein-specific scFv were detected only in those plants transformed with the expression cassettes containing the coding sequence for the PHA signal peptide (Figure 10.1b-3). This signal peptide was previously reported to direct the secretion of pea seed albumin into the culture fluid. Our experiments with cell suspension cultures indicated that the BNYVV coat protein-specific scFv were retained in the cells, presumably in the ER or an ER-associated compartment. Since our constructs did not code for the KDEL motif responsible for the retention of proteins in the ER (Fiedler and Conrad, 1995; Schouten *et al.*, 1996), it is possible that some internal amino acid motifs within the BNYVV coat protein-specific scFv might have caused their retention in the cells. Others have observed that scFv constructs which included a secretory signal sequence but no C-terminal KDEL ER retention signal have resulted in the protein product being retained within the cells (for example, Richardson *et al.*, 1995, and references therein).

The expression of BNYVV coat protein-specific scFv had no obvious effects on the growth or the phenotype of the transformed plants as long as levels of expression were 0.01 per cent or less of the TSP. Selfing of some of the plants led to increased expression levels up to 0.1 per cent. Such plants showed some growth retardation. The expression of scFv specific for the 25K protein of BNYVV had apparently a much more pronounced effect on the metabolism of the transformed plants. They grew only poorly and failed to form roots on kanamycin-free media. This suggests that the BNYVV 25K protein-specific scFv may interfere with some normal host constituent with a structural and possibly also functional similarity to the 25K protein, possibly a factor which is necessary for root formation. In BNYVV-infected plants the abundant presence of the 25K protein is apparently the cause for the excessive root formation which has led to the name 'rhizomania'. The blockage of small amounts of a 25K protein-related factor in healthy plants could possibly prevent root formation altogether. In the dot blot immunoassay we could not detect any reactivity of our 25K protein-specific Mab or scFv with extracts from healthy plants. Some cross-reactivity was observed however, in ELISA on plates precoated with polyclonal antibodies. In this type of test, discontinuous epitopes are more likely to be detected.

Although the BNYVV coat protein-specific scFv which we could produce in the plants were targeted to the secretory pathway and accumulated in the ER or an ER-associated compartment rather than in the cytoplasm, they were apparently able to interfere with both early and late stages of BNYVV infection. The inhibition of the establishment of BNYVV in the plants early in infection might be due to a blockage of the virus which in disrupted cells may come into contact with scFv leaking from the damaged ER. This blocking effect of the scFv would be expected to be counteracted with a more concentrated inoculum. This was indeed observed in our experiments. Once the virus became established in transgenic line 23 there was no inhibition of virus multiplication. This multiplication presumably takes place in the cytoplasm, which in the healthy cells of our transgenic plants probably contained no scFv. The somewhat lower average virus concentrations observed with transgenic line 116 may possibly have been due to the less favourable conditions for virus multiplication offered by the growth-retarded plants.

BNYVV produces visible pathogenic effects on *N. benthamiana*, mainly on old plants after a long period of infection. The coat protein-specific scFv exerted a protective effect against these late pathogenic effects. It is known that the coat proteins of several plant viruses belonging to various taxonomic groups, for example, of TMV (Banerjee *et al.*, 1995; Culver and Dawson, 1989), *Potato virus X* (genus *Potexvirus*) (Santa Cruz and Baulcombe, 1993), *Alfalfa mosaic virus* (genus *Alfamovirus*) (Neeleman *et al.*, 1991), *Cucumber mosaic virus* (genus *Cucumovirus*) (Suzuki *et al.*, 1995) and *Turnip crinkle virus* (genus *Carmovirus*) (Heaton and Laakso, 1995) are involved in pathogenicity. The coat protein-specific scFv may block these coat protein-mediated virus–host interactions for which the mechanisms are not yet understood. The diminished pathogenicity of BNYVV for the scFv-expressing plants in the late stages of infection may thus be due to interference of the scFv with coat protein-mediated virus–host interactions.

Acknowledgements

The authors are greatly indebted to the Deutsche Forschungsgemeinschaft (grant Ko518/12-1) and the European Union (AIR3 project CT94-1046) for financial support.

References

AN, G., EBERT, P.R., MITRA, A. and HA, S.B., 1988, Binary vectors, in *Plant Molecular Biology Manual*, edited by S.B. Gelvin, R.A. Schilperoort and D.P.S. Verma, Dordrecht: Kluwer Academic Publishers, pp. A3/7–A3/8.

BANERJEE, N., WANG, J.Y. and ZAITLIN, M., 1995, A single nucleotide change in the coat protein gene of tobacco mosaic virus is involved in the induction of severe chlorosis, *Virology*, **207**, 234–239.

BENVENUTO, E., ORDAS, R.J., TAVAZZA, R., ANCORA, G., BIOCCA, S., CATTANEO, A. and GALEFFI, P., 1991, 'Phytoantibodies': a general vector for the expression of immunoglobulin domains in transgenic plants, *Plant Molecular Biology*, **17**, 865–874.

BIOCCA, S., RUBERTI, F., TAFANI, M., PIERANDREI-AMALDI, P. and CATTANEO, A., 1995, Redox state of single chain Fv fragments targeted to the ER, cytosol and mitochondria, *Biotechnology*, **13**, 1110–1115.

CLARK, M. and ADAMS, A.M., 1977, Characteristics of the microplate method of enzyme-linked immunosorbent assay for the detection of plant viruses, *Journal of General Virology*, **34**, 475–483.

COMMANDEUR, U., KOENIG, R., MANTEUFFEL, R., TORRANCE, L., LÜDDECKE, P. and FRANK, R., 1994, Location, size and complexity of epitopes on the coat protein of beet necrotic yellow vein virus studied by means of synthetic overlapping peptides, *Virology*, **198**, 282–287.

CULVER, J.N. and DAWSON, W.O., 1989, Point mutations in the coat protein gene of tobacco mosaic virus induce hypersensitivity in *Nicotiana sylvestris*, *Molecular Plant Microbe Interaction*, **2**, 209–213.

DOREL, C., VOELKER, T.A., HERMAN, E.M. and CHRISPEELS, M.J., 1989, Transport of proteins to the plant vacuole is not by bulk flow through the secretory system, and requires positive sorting information, *Journal of Cell Biology*, **108**, 327–337.

ENGELHARDT, O., GRABHERR, R., HIMMLER, G. and RUEKER, F., 1994, Two-step cloning of antibody variable domains in a phage display vector – Fv single chain antibody engineering using phagemid pCOCK for light chain and heavy chain variable region gene cloning, *BioTechniques*, **17**, 44–46.

FARINELLI, L., MALNOE, P. and COLLET, G.F., 1992, Heterologous encapsidation of potato virus Y strain O (PVY^O) with the transgenic coat protein of PVY strain N (PVY^N) in *Solanum tuberosum* cv. Bintje, *Bio/Technology*, **10**, 1020–1025.

FECKER, L.F., KAUFMANN, A., COMMANDEUR, U., COMMANDEUR, J., KOENIG, R. and BURGERMEISTER, W., 1996, Expression of single chain antibody fragments (scFv) specific for beet necrotic yellow vein virus coat protein or 25 kDa protein in *Escherichia coli* and *Nicotiana benthamiana*, *Plant Molecular Biology*, **32**, 979–986.

FECKER, L.F., KOENIG, R. and OBERMEIER, C., 1997, *Nicotiana benthamiana* plants expressing beet necrotic yellow vein virus (BNYVV) coat protein-specific scFv are partially protected against the establishment of the virus in the early stages of infection and its pathogenic effects in the late stages of infection, *Archives of Virology*, **142**, 1857–1863.

FIEDLER, U. and CONRAD, U., 1995, High level production and long-term storage of engineered antibodies in transgenic tobacco seeds, *Bio/Technology*, **13**, 1090–1093.

GALLIE, D.R., SLEAT, D.E., WATTS, J.W., TURNER, P.C. and WILSON, T.M.A, 1987, The 5′-leader sequence of tobacco mosaic virus RNA enhances the expression of foreign gene transcripts *in vitro* and *in vivo*, *Nucleic Acids Research*, **15**, 3257–3273.

GREENE, A.E. and ALLISON, R.F., 1994, Recombination between viral RNA and transgenic plant transcripts, *Science*, **263**, 1423–1425.

HEATON, L.A. and LAAKSO, M.M., 1995, Several symptom-modulating mutations in the coat protein of turnip crinkle carmovirus result in particles with aberrant conformational properties, *Journal of General Virology*, **76**, 225–230.

HOEKEMA, A., HIRSCH, P.R., HOOYKAAS, P.J.J. and SCHILPEROORT, R.A., 1983, A binary plant vector strategy based on separation of vir- and T-region of the *Agrobacterium tumefaciens* Ti-plasmid, *Nature*, **303**, 179–180.

HORSCH, R.B., FRY, J.E., HOFFMANN, N.L., EICHHOLTZ, D., ROGERS, S.G. and FRALEY, R.T., 1985, A simple and general method for transferring genes into plants, *Science*, **227**, 1229–1231.

HUNT, D.C. and CHRISPEELS, M.J., 1991, The signal peptide of a vacuolar protein is necessary and sufficient for the efficient secretion of a cytosolic protein, *Plant Physiology*, **96**, 18–25.

KABAT, E.A., WU, T.T., PERRY, H.M., GOTTESMAN, K.S. and FOELLER, C., 1991, *Sequences of Proteins of Immunological Interest*, 5th edition, US Dept of Health and Human Services, Washington DC.

KAUFMANN, A., 1995, Molekularbiologische und serologische Untersuchungen über das beet necrotic yellow vein virus (BNYVV) und das beet soil borne virus (BSBV) – Erarbeitung der Grundlagen für eine gentechnisch zu induzierende immunologische Resistenz gegen das BNYVV, PhD thesis, University Braunschweig, Germany, ISBN 3-930697-58-0.

KOENIG, R. and LESEMANN, D.-E., 1999, *Genus Benyvirus*, *Virus Taxonomy: Seventh Report of the International Committee on Taxonomy of Viruses*, Eds M.H.V. Van Regenmortel *et al.*, Academic Press (in press).

KOENIG, R., JARAUSCH, W., LI, Y., COMMANDEUR, U., BURGERMEISTER, W., GEHRKE, M. and LÜDDECKE, P., 1992, Effect of recombinant beet necrotic yellow vein virus with different RNA compositions on mechanically inoculated sugarbeets, *Journal of General Virology*, **72**, 2243–2246.

LANDSMANN, J., LLEWELLYN, D., DENNIS, E.S. and PEACOCK, W.J., 1988, Organ regulated expression of the *Parasponia andersonii* haemoglobin gene in transgenic plants, *Molecular General Genetics*, **214**, 68–73.

LEI, S.-P., LIN, H.-C., WANG, S.-S., CALLAWAY, J. and WILCOX, G., 1987, Characterization of the *Erwinia carotivora* pelB gene and its product pectate lyase, *Journal of Bacteriology*, **169**, 4379–4383.

NEELEMAN, L., VAN DER KUYL, A.C. and BOL, J.F., 1991, Role of alfalfa mosaic virus coat protein gene in symptom formation, *Virology*, **181**, 687–693.

OW, D.W., JACOBS, J.D. and HOWELL, S.H., 1987, Functional regions of the cauliflower mosaic virus 35S RNA promoter determined by use of the firefly luciferase gene as a reporter of promoter activity, *Proceedings of the National Academy of Sciences USA*, **84**, 4870–4874.

PETERS, D. and GODFREY-VELTMAN, A., 1989, Polymyxa betae zoospores as vectors of beet necrotic yellow vein furovirus, *Bull OEPP/EPPO Bull*, **19**, 509–515.

RICHARDS, K. and TAMADA, T., 1992, Mapping functions on the multipartite genome of beet necrotic yellow vein virus, *Annual Review of Phytopathology*, **30**, 291–313.

RICHARDSON, J.H., SODROSKI, J.G., WALDMANN, T.A. and MARASCO, W.A., 1995, Phenotypic knockout of the high-affinity human interleukin 2 receptor by intracellular single-chain antibodies against the α subunit of the receptor, *Proceedings of the National Academy of Sciences USA*, **92**, 3137–3141.

SANTA CRUZ, S. and BAULCOMBE, D., 1993, Molecular analysis of potato virus X isolates in relation to the potato hypersensitivity gene Nx, *Molecular Plant–Microbe Interaction*, **6**, 707–714.

SAUNAL, H. and VAN REGENMORTEL, M.H.V., 1995, Kinetic and functional mapping of viral epitopes using biosensor technology, *Virology*, **213**, 462–471.

SAUNAL, H., WITZ, J. and VAN REGENMORTEL, M.H.V., 1993, Inhibition of *in vitro* cotranslational disassembly of tobacco mosaic virus by monoclonal antibodies to the viral coat protein, *Journal of General Virology*, **74**, 897–900.

SCHOUTEN, A., ROOSIEN, J., VAN ENGELEN, F.A., DE JONG, G.A., BORST-VRENSSEN, A.W., ZILVERENTANT, J.F., *et al.*, 1996, The C-terminal KDEL sequence increases the expression level of a single-chain antibody designed to be targeted to both the cytosol and the secretory pathway in transgenic tobacco, *Plant Molecular Biology*, **30**, 781–793.

SUZUKI, M., KUWATA, S., MASUTA, C. and TAKANAMI, Y., 1995, Point mutations in the coat protein of cucumber mosaic virus affect symptom expression and virion accumulation in tobacco, *Journal of General Virology*, **76**, 1791–1799.

TAMADA, T., SAITO, M., KIGUCHI, T. and KUSUME, T., 1990, Effect of isolates of beet necrotic yellow vein virus with different RNA components on the development of rhizomania symptoms; in *Schriftenreihe Deutsche Phytomedizinische Gesellschaft*, Bd 1, Proc 1st Symp IWGPVFV, Braunschweig, Aug 1990, Compiled by R. Koenig, Ulmer Verlag, Stuttgart 1, 41–44.

TAVLADORAKI, P., BENVENUTO, E., TRINCA, S., DE MARTINIS, D., CATTANEO, A. and GALEFFI, P., 1993, Transgenic plants expressing a functional single-chain Fv antibody are specifically protected from virus attack, *Nature*, **366**, 469–472.

TEPFER, M., 1993, Viral genes and transgenic plants. What are the potential risks?, *Bio/Technology*, **11**, 1125–1132.

TÖPFER, R., MATZEIT, V., GRONENBORN, B., SCHELL, J. and STEINBISS, H., 1987, A set of plant expression vectors for transcriptional and translational fusions, *Nucleic Acids Research*, **15**, 5890.

VOSS, A., NIERSBACH, M., HAIN, R., HIRSCH, H.J., LIAO, Y.C., KREUZALER, F. and FISCHER, R., 1995, Reduced virus infectivity in *N. tabacum* secreting a TMV-specific full size antibody, *Molecular Breeding*, **1**, 39–50.

WHITELAM, G., COCKBURN, B. and OWEN, M.R.L., 1994, Antibody production in transgenic plants, *Biochemical Society Transactions*, **22**, 940–944.

WINTERMANTEL, W.M. and SCHOELZ, J.E., 1996, Isolation of recombinant viruses between cauliflower mosaic virus and a viral gene in transgenic plants under conditions of moderate selection pressure, *Virology*, **223**, 156–164.

11

Engineering Antibody-mediated Resistance to Plant Root Nematodes – Potential and Prospects

GEERT SMANT, JAAP BAKKER, FRED J. GOMMERS AND ARJEN SCHOTS

11.1 Plant root nematodes and parasitism

Plant-parasitic nematodes have developed diverse adaptations enabling them to parasitize roots. This has involved the development of a protrusible stylet (to penetrate cell walls and feed on cell contents) and marked morphological and physiological adaptations of the oesophagus. Cells of the oesophagus gland enlarged and the nature and function of their secretions changed as nematodes adapted to parasitism. The majority of the physiological functions of these secretions are as yet, not fully understood. However, their roles are thought to be of a pivotal nature.

Nematode parasitism involves complex and dynamic interactions between the nematode and the host plant. These include hatching stimuli, attraction to the host, penetration of host tissue, recognition of tissues suitable for feeding site formation, modification of host tissue and an active response from the host plant.

Sedentary endoparasitic cyst (*Globodera* spp. and *Heterodera* spp.) and root knot (*Meloidogyne* spp.) nematodes have developed a highly specialized manner of parasitizing host roots. After root penetration they migrate to the stele where they induce a feeding site. At the cytological level, the route taken by cyst nematodes, the cells chosen for syncytium formation (see Figure 11.1) and also the ultrastructural changes that take place during syncytium development (Golinowski *et al.*, 1996) are known. The most intriguing phenomenon is the way sedentary endoparasitic nematodes manipulate their host cells such that the cell is made into a feeding cell. Insight is emerging into the molecular basis of this part of the pathogenesis process (for an overview see Fenoll *et al.*, 1997) such as the re-entry of young feeding cells into the cell-cycle (Niebel *et al.*, 1996) and the increase in turgor pressure in these cells. Genes such as the *Lea*-like gene, which encode for proteins thought to be involved in osmoprotection, and a tobacco root-specific gene encoding a putative water channel have been cloned (Opperman *et al.*, 1994; Van der Eycken *et al.*, 1996). Also, an increase in extensin expression has been observed. This protein may have a role in fortifying the cell wall (Niebel *et al.*, 1993). In addition to proteins whose expression levels are upregulated in feeding cells, several pathogenesis related proteins have been found in other tissues (Niebel *et al.*, 1995).

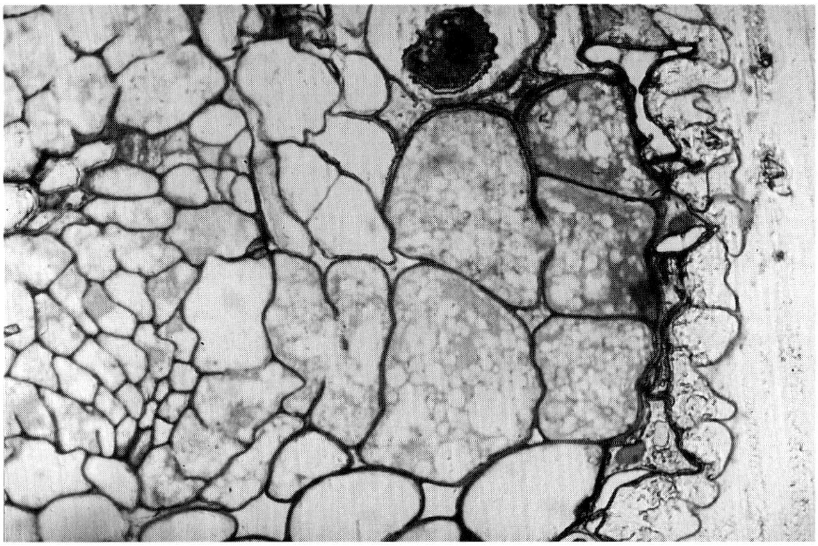

Figure 11.1 Section through a potato root (cultivar Bintje) infected with *Globodera rostochiensis*. The syncytium is clearly visible. It contains many small vacuoles and the cell wall between cells is dissolved in several places.

During parasitism the functions of the oesophageal glands change. The subventral glands are assumed to be active mainly in the initial stages of parasitism, during root penetration, root migration and feeding cell induction. These glands are packed with secretory granules in preparasitic and parasitic second stage juveniles, but not in the later stages of nematode development. In contrast, the dorsal gland contains only a few granules in the earlier stages and fills up with secretory granules at the end of migration and during the initial stages of feeding cell induction. It is generally accepted that the compounds used to form a feeding tube arise from the dorsal gland. Feeding tubes which function as molecular sieves are formed prior to withdrawal of cell contents. It is, as yet, unclear what the functions of the subventral glands are, once the nematode has become immobile, or which other functions the dorsal gland has during this stage.

11.2 Antibodies: a new plant resistance factor

Knowledge of the molecular basis of nematode parasitism can be used to design antibody-mediated resistance traits. Antibodies offer a good methodology for obtaining resistance. They can have a profound effect on the biological activity of proteins upon binding (for an overview, see Schots *et al.*, 1992). This is caused by either steric hindrance, or by a forced change in the three-dimensional structure of the protein, with a consequent loss of function. This was shown *in vitro* for several proteins as well as *in vivo* in animal cells (see, for example, Rungger *et al.*, 1979; Feramisco *et al.*, 1985; Riabowol *et al.*, 1988) and in plant cells upon heterologous expression of antibody genes. It was shown that the expression of antibodies *in planta* inhibits replication of *Artichoke mottled crinkle virus* (AMCV, genus *Tombusvirus*) (Tavladoraki *et al.*, 1993) and *Tobacco mosaic virus* (TMV, genus *Tobamovirus*) (Voss *et al.*, 1995). Physiological processes may also be influenced. Plant expression of anti-phytochrome antibodies resulted in aberrant phytochrome-dependent germination (Owen *et al.*, 1992). Seed-specific immunomodulation of abscisic acid (ABA) by an anti-ABA antibody resulted in a switch from the seed maturation programme to a germination programme (Phillips *et al.*, 1997; see also Chapter 9, this book). These examples clearly demonstrate the feasibility of immunomodulation in plants.

11.2.1 Isolation and characterization of targets

The implementation of antibody-mediated resistance against sedentary plant parasitic nematodes is only possible if suitable targets are available. However, the molecular basis of nematode parasitism is poorly understood. This is a consequence of the small size and the obligate nature of these soil pathogens. Recent advances in biochemistry and molecular biology have enabled progress in this field. The understanding of plant responses to nematode infestation is increasing (Fenoll *et al.*, 1997). In contrast, little is known about those molecules produced by the nematode which allow it to parasitise host roots. Only one has recently been characterized and is described below. As a consequence, it is necessary to isolate, possibly clone and characterize more of these nematode molecules prior to raising inhibitory antibodies for *in planta* expression.

The isolation of target molecules was approached by raising monoclonal antibodies against the contents of the oesophageal glands (Davis *et al.*, 1992, 1994; Goverse *et al.*, 1994; Hussey *et al.*, 1994; De Boer *et al.*, 1996a, 1996b). Shotgun approaches were initially used. Crude nematode homogenates were used to immunize mice, and monoclonal antibodies were selected after a hybridoma cell fusion. Selections took place using an ingenious protocol based on immunofluorescence on fixed tissue sections of sliced nematodes (Atkinson *et al.*, 1988; Hussey, 1989). This resulted in the production of monoclonal antibodies specific for the subventral or dorsal glands of *Heterodera glycines, Globodera rostochiensis* and *Meloidogyne* sp. Further characterization showed that most of these monoclonal antibodies bind to structural proteins (Ray *et al.*, 1994). It was not possible to prove that the others bind secretory proteins (De Boer *et al.*, 1996a) within the gland cells. Goverse *et al.* (1994) isolated three monoclonal antibodies that bound consistently to the subventral gland secretions of *H. glycines*.

Hence, new strategies were developed to raise monoclonal antibodies that bind to the secretory proteins of sedentary endoparasites. Davis *et al.* (1992) used homogenates of the anterior regions and stylet secretions of *M. incognita* adult females as the immunogen. Nine monoclonal antibodies were produced that bind to the secretory granules in the dorsal gland, and two that bind to the secretory granules within the subventral gland. However, attempts to use these antibodies to isolate secretory proteins from the gland by immunoaffinity chromatography failed (Hussey and Davis, personal communication).

In our laboratory we decided to use an alternative strategy based on the partial purification of nematode proteins (De Boer *et al.*, 1996b). Homogenized second stage juveniles of *G. rostochiensis* were separated into 50 protein fractions by preparative continuous flow electrophoresis in the presence of sodium dodecyl sulphate (SDS). The molecular mass of these proteins ranged from 30 kDa to 52 kDa. After examination of the individual fractions, the fractions were pooled, giving a set of 16 samples of increasing molecular mass. These were used to immunize 16 individual mice. After the second immunization, antisera were collected and screened by immunofluorescence for binding to the oesophageal glands of second stage juveniles of *G. rostochiensis*. The sera of the eight mice immunized with the successive protein fractions from 36 kDa to 52 kDa showed a strong reaction with the subventral oesophageal gland cells. One mouse, immunized with the 38 kDa to

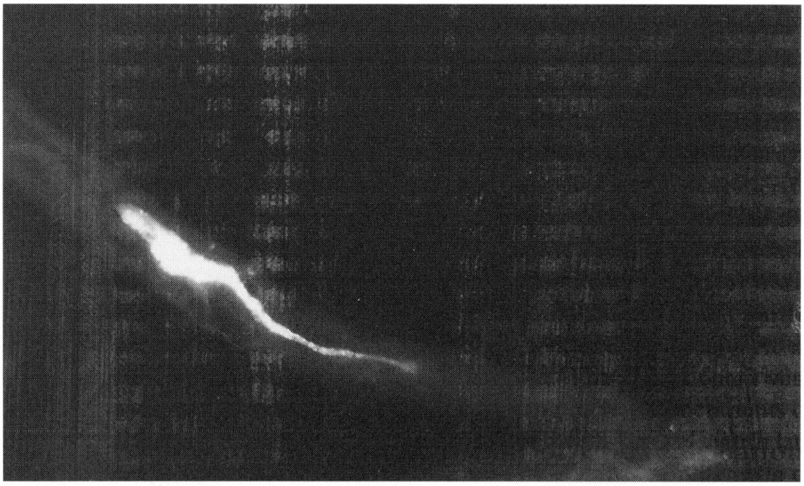

Figure 11.2 Immunofluorescent staining of a sliced *Globodera rostochiensis* juvenile with monoclonal antibody MGR 48. The subventral oesophageal gland and its extension is clearly visible.

40.5 kDa protein sample that was chosen for monoclonal antibody production. Using immunofluorescence microscopy, ten monoclonal antibodies were selected which bound to the entire contents of the subventral gland cells including the gland extensions (see Figure 11.2). Western blotting revealed that these monoclonal antibodies bound one or more of four subventral gland proteins (svp) with a molecular mass of 30 kDa (svp30), 31 kDa (svp31), 39 kDa (svp39) and 49 kDa (svp49).

Further characterization of these clones showed that they almost exclusively bound to members of the genus *Globodera*. Only monoclonal antibody MGR 48, which binds to all four svps, showed cross-reactivity by binding to *Heterodera glycines* and *H. schachtii*. Electron microscopy using MGR 48 as the probe revealed that the protein antigens localize mainly in the electron dense core of the secretory granules (Figure 11.3), indicating that the svps are indeed secreted.

The subventral glands terminate in ampullae within the oesophagus, located posterior to the metacorporal pump chamber. It has been argued that the metacorporal pump chamber acts as a unidirectional valve that precludes forward flow of secretions (Wyss, 1992). As a consequence subventral gland secretions can be transported only posteriorly, and so are thought to be involved in digestion or mobilization of lipid reserves. Recently, we have been able to show that the opposite is true (Smant *et al.*, 1997). When second stage juveniles of *G. rostochiensis* were hatched in potato root diffusate, and after extensive washing were incubated in standard pore water (SPW; a water solution mimicking ground water) the antigens were secreted into the SPW (see Figure 11.4). Juveniles that hatch spontaneously in tap water do not secrete these proteins. Western blot analysis revealed three of the svps (31 kDa, 39 kDa and 49 kDa) to be in SPW, with svp31 predominating.

Developmental regulation of svps throughout the nematode's life cycle was determined by Western blot analysis. Maximum levels of svps were detected in homogenates of preparasitic hatched juveniles. The levels decreased in parasitic second stage juveniles. In later developmental stages the svps could not be detected except for low levels of svp 49 in adult males. To obtain evidence that svps are secreted *in planta* during the initial

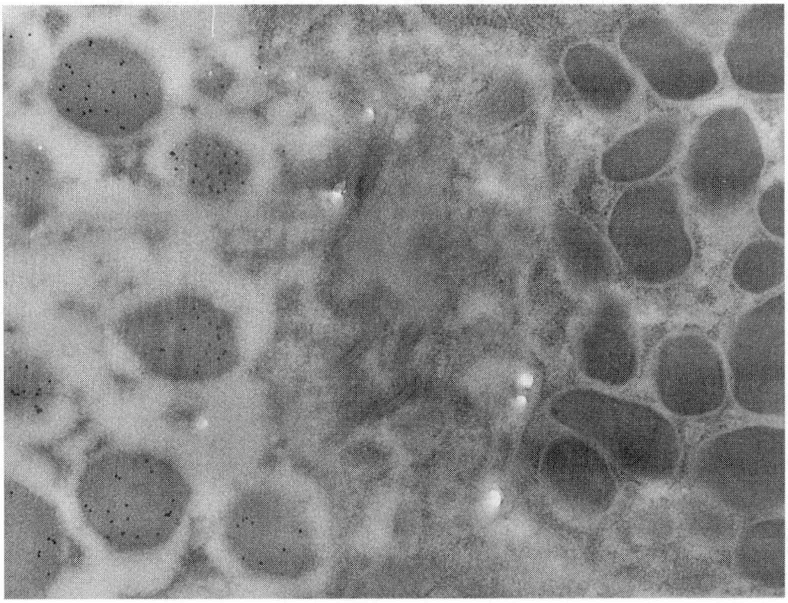

Figure 11.3 Micrograph of the oesophageal glands of a second stage juvenile of *Globodera rostochiensis* labelled with monoclonal antibody MGR 48 specific for the subventral gland and visualized with an anti-mouse immunogold conjugate. The contents of the secretory vesicles of the subventral gland are labelled, while the more electron dense granules of the dorsal gland remain unlabelled.

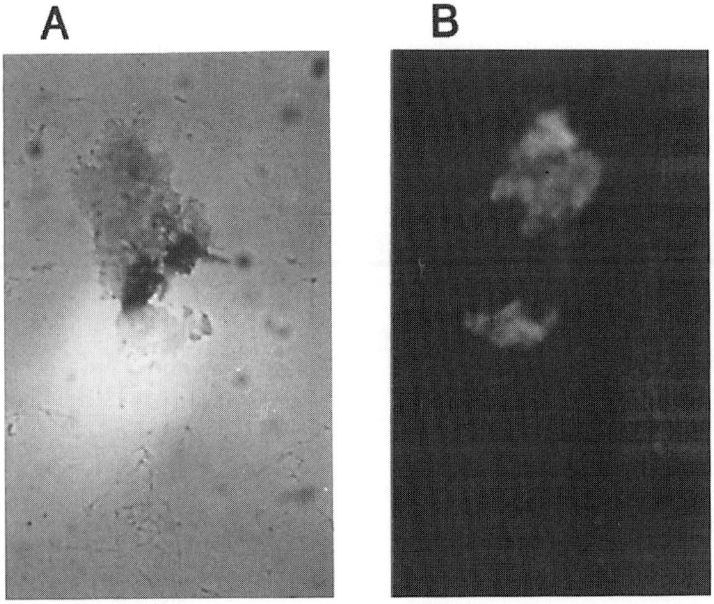

Figure 11.4 (A) Second stage juveniles of *Globodera rostochiensis* incubated in potato root diffusate secrete the contents of the oesophageal glands. Under acid conditions the secretions precipitate. (B) Immunofluorescence staining of these precipitates incubated with monoclonal antibodies specific for the subventral oesophageal gland (staining with MGR 48).

stages of parasitism, ultrathin sections of fixed parasitized roots were probed with MGR 48 conjugated to immunogold. It was impossible to detect the presence of svps in plant tissue. However, their presence was assessed in the lumen of the oesophagus of parasitic second stage juveniles.

11.2.2 Cloning of targets to reveal their function and site of expression

Subventral gland proteins are actively secreted from the subventral oesophageal glands of *G. rostochiensis*. Hence, a role in pathogenesis seems plausible. To assess the function of these proteins we decided to clone their genes (Smant *et al.*, 1998). Protein homogenates of *G. rostochiensis* and *H. glycines* were separated by preparative continuous flow electrophoresis. Those fractions showing a reaction with MGR 48 on a Western blot were used for further protein purification by immunoaffinity chromatography (Figure 11.5). The N-terminal amino acid sequences of the purified proteins were then determined. Sequences were obtained for svp39 of *G. rostochiensis* and svp49 of *H. glycines*. Based on these sequences, primers were designed and used to amplify the genes from cDNA libraries by PCR. Two PCR products were obtained for each of the two nematode species. These genes were cloned and their DNA sequences determined (see Table 11.1). The four encoded proteins were compared with sequences currently available in protein databases. The search indicated that the nematode proteins have significant homologies to β-1,4-endoglucanases (EGases) of various bacterial origins. The proteins GR-ENG-1 and HG-ENG-1 are comprised of distinct domains. Each has a cellulolytic domain at the N-terminus connected by a typical linker sequence to a type II cellulose binding domain at the C-terminus.

The question that remained was whether or not the cloned genes indeed encode β-1,4-endoglucanases. This was investigated using an EGase activity assay. Proteins from

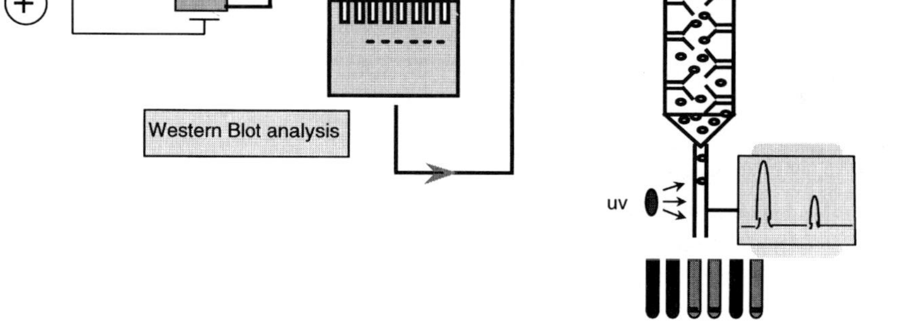

Figure 11.5 Schematic representation of the method employed to purify subventral gland antigens. After preparative electrophoresis the antigen-containing fractions were identified by Western blotting. These fractions were further purified using immuno-affinity chromatography.

Table 11.1 Summary of the cDNA clones obtained, their sizes and molecular mass of the putative protein

Species	Clone	cDNA	Open reading frame in clone	Molecular mass of putative protein (Da)
G. rostochiensis	GR-*eng-1*	1546	1416	49 700
G. rostochiensis	GR-*eng-2*	1300	1179	42 000
H. glycines	HG-*eng-1*	1615	1428	49 800
H. glycines	HG-*eng-2*	1191	957	34 700

homogenates were separated by SDS polyacrylamide gel electrophoresis using a lower SDS concentration. Enzyme activity was then assessed by overlaying the gel on an agarose bed containing carboxymethylcellulose (CMC). In the *G. rostochiensis* homogenate a protein of approximately 50 kDa showed CMC hydrolase activity, whereas in *H. glycines* a slightly smaller protein hydrolysed CMC. Both apparent EGases aligned with one of the subventral gland specific antigens of approximately 49 kDa recognized by MGR 48.

To localize EGase transcripts, whole mount *in situ* hybridization experiments were done. Anti-sense RNA probes transcribed from HG-*eng-2* cDNA bound specifically within the subventral glands of preparasitic second stage juveniles of *H. glycines*.

11.3 Future prospects

Cyst nematodes migrate intracellularly towards the stele. A procambium cell is then selected as the initial syncytial cell (Golinowski *et al.*, 1996). Originally, the intracellular migration was thought to be facilitated by mechanical means only. However, the cloning of β-1,4-endoglucanase indicates that the intracellular migration through the cell walls of host roots involves a combination of mechanical piercing and enzymatic softening of the cell walls. The β-1,4-endoglucanases from *G. rostochiensis* and *H. glycines* are the first pathogenicity factors to be cloned from a plant parasitic nematode. In addition to cellulose as a possible substrate, EGases may also be involved in the hydrolysis of xyloglucan (the major component of hemicellulose). This polysaccharide interconnects the crystalline cellulose microfibrils in type I primary cell walls. Because some of the EGases described contain a cellulose binding domain, it seems likely that these enzymes allow the nematode to weaken the strength of the cellulose microfibril network in the plant cell wall.

In addition to cellulose and xyloglucan, the plant cell wall consists of substantial amounts of pectin, pectin-like structures and other proteins. It is therefore likely that sedentary endoparasitic nematodes also produce other cell wall degrading enzymes such as pectinases and proteases.

Since sedentary endoparasitic nematodes use enzymes as well as purely mechanical means for their migration within host roots, inhibition of these enzymes would be likely to hinder them reaching their destination. Juveniles of these nematodes are not able to take up food before induction of a feeding site. They rely completely on the lipid reserves in their intestines which provide the energy source for migration through the soil towards the host root, subsequent penetration and migration. As stated before, enzymes can be efficiently inhibited by antibodies. When raising a panel of monoclonal antibodies against an enzyme (by traditional hybridoma technology or by selection from antibody phage display libraries) usually a large percentage of antibodies will substantially inhibit the enzyme. Therefore, antibodies may be well suited to inhibiting cell wall degrading enzymes.

The technology to express antibodies in the apoplast is well developed. Full size antibodies can be expressed without difficulty to high levels equalling 1 per cent or more of total root protein (Van Engelen *et al.*, 1994). Therefore, *in planta* expression of antibodies may be useful for obtaining resistance against intercellularly migrating nematodes such as *Meloidogyne* spp. and intracellularly migrating nematodes such as cyst nematodes. In addition, intracellular expression levels of single-chain antibody fragments can also reach levels of 1 per cent or higher (Schouten *et al.*, 1996). Such symplastic expression may also be useful for intracellularly migrating nematodes such as *Heterodera* spp. and *Globodera* spp. as they can directly bind to proteins secreted from the oesophageal glands during parasitism. In addition, those molecules secreted by sedentary nematodes for inducing or maintaining a feeding site may be targeted by antibodies expressed intracellularly.

The expression of antibodies against an oesophageal gland protein of *M. incognita* in plant cells has been tried. Rosso *et al.* (1996) constructed an scFv from the IgM monoclonal antibody 6D4 (Hussey *et al.*, 1990). This antibody recognizes a high molecular weight glycoprotein (> 212 kDa). The scFv was stable upon transient expression in tobacco protoplasts. Independent *Agrobacterium*-mediated transformation of tobacco with the 6D4 heavy and light chain genes resulted in transgenic plants producing either the heavy or the light chain (Baum *et al.*, 1996). These plants were crossed resulting in progeny producing functional antibodies as revealed in ELISA and immunofluorescence. Surprisingly, non-functional heavy chain dimers were also produced which were not sensitive to proteolytic breakdown. However, the molecular weights of the heavy chains produced by the plants are smaller than those produced in hybridomas. The antibodies did not affect root-knot nematode parasitism. Either the antibody had no effect on any actions of the antigen, or this glycoprotein has no pivotal role in parasitism.

Research concerning antibody-mediated resistance against plant parasitic nematodes has so far focused on sedentary endoparasites. The question is whether or not this approach can be applied to obtain resistance against other nematodes as well. Migratory endoparasites such as *Pratylenchus* spp. and *Radopholus* spp. penetrate host roots and disrupt a large number of root cells. It has to be expected that these nematodes also use cell wall degrading enzymes. Hence, suitable targets may be available. Migratory ectoparasites which feed on cells below the epidermis are a third category for engineering this type of resistance. The Dorylaimid genera *Xiphinema* and *Longidorus* and Tylenchid *Helicotylenchus* spp. belong to this group. Many of these species also invoke a kind of feeding cell often accompanied by typical swelling, due to a hyperplastic response around the penetration site of the root (Sijmons *et al.*, 1994). Furthermore, these nematodes often feed for prolonged periods. For instance, *Xiphinema* spp. induce so called coenocytes. These are cells that are formed by repeated mitosis without cell division. Feeding on these cells can last up to two weeks. These two groups exemplify the idea that antibodies may be used to engineer resistance against a wide range of nematodes which use different strategies to parasitize host roots.

11.4 Conclusions

In conclusion, the 'plantibody' strategy is a valuable method for obtaining transgenic resistance against various groups of nematodes. The fact that many species are polyphagous, hinders the application of natural resistance. This makes the use of antibodies a worthwhile alternative. High level expression in plants, and the possibility of inhibiting the function of proteins and other compounds involved in nematode pathogenesis render it a promising approach for obtaining resistance.

References

ATKINSON, H.J., HARRIS, P.D., HALK, E.J., NOVITSKI, C., LEIGHTON-SANDS, J., NOLAN, P. and FOX, P.C., 1988, Monoclonal antibodies to the soybean cyst nematode, *Heterodera glycines, Annals of Applied Biology*, **112**, 459–469.

BAUM, T.J., HIATT, A., PARROT, W.A., PRATT, L.H. and HUSSEY, R.S., 1996, Expression in tobacco of a functional monoclonal antibody specific to stylet secretions of the root-knot nematode, *Molecular Plant–Microbe Interactions*, **9**, 382–387.

DAVIS, E.L., ARON, L.M., PRATT, L.H. and HUSSEY, R.S., 1992, Novel immunization procedures to develop monoclonal antibodies that bind to specific structures in *Meloidogyne* spp., *Phytopathology*, **82**, 1244–1250.

DAVIS, E.L., ALLEN, R. and HUSSEY, R.S., 1994, Developmental expression of esophageal gland antigens and their detection in stylet secretions of *Meloidogyne incognita, Fundamental and Applied Nematology*, **17**, 255–262.

DE BOER, J.M., OVERMARS, H.A., POMP, H., DAVIS, E.L., ZILVERENTANT, J.F., GOVERSE, A., *et al.*, 1996a, Production and characterization of monoclonal antibodies to antigens from second stage juveniles of the potato cyst nematode, *Globodera rostochiensis, Fundamental and Applied Nematology*, **19**, 545–554.

DE BOER, J.M., SMANT, G., GOVERSE, A., DAVIS, E.L., OVERMARS, H.A., POMP, H., *et al.*, 1996b, Secretory granule proteins from the subventral esophageal glands of the potato cyst nematode identified by monoclonal antibodies to a protein fraction from second-stage juveniles, *Molecular Plant–Microbe Interactions*, **9**, 39–46.

FENOLL, C., GRUNDLER, F.M.W. and OHL, S.A., 1997, *Cellular and Molecular Aspects of Plant–Nematode Interactions*, Dordrecht: Kluwer Academic Publishers.

FERAMISCO, J.R., CLARK, R., WONG, G., ARNHEIM, N., MILLEY, R. and MCCORMICK, F., 1985, Transient reversion of ras oncogene-induced cell transformation by antibodies specific for amino acid 12 of ras protein, *Nature*, **314**, 639–642.

GOLINOWSKI, W., GRUNDLER, F.M.W. and SOBCZAK, M., 1996, Changes in the structure of *Arabidopsis thaliana* during female development of the plant-parasitic nematode *Heterodera schachtii, Protoplasma*, **194**, 103–116.

GOVERSE, A., DAVIS, E.L. and HUSSEY, R.S., 1994, Monoclonal antibodies to the esophageal glands and stylet secretions of *Heterodera glycines, Journal of Nematology*, **26**, 251–259.

HUSSEY, R.S., 1989, Monoclonal antibodies to secretory granules in esophageal glands of *Meloidogyne* species, *Journal of Nematology*, **21**, 392–398.

HUSSEY, R.S., PAGUIO, O.R. and SEABURY, F., 1990, Localization and purification of a secretory protein from the esophageal glands of *Meloidogyne incognita* with a monoclonal antibody, *Phytopathology*, **80**, 709–714.

HUSSEY, R.S., DAVIS, E.L. and RAY, C., 1994, *Meloidogyne* stylet secretions, in Lamberti, F., De Giorgi, C. and Bird, D.M. (Eds), *Advances in Molecular Plant Nematology*, NATO ASI Series, Series A. Life Sciences, pp. 233–249, New York: Plenum Press.

NIEBEL, A., DE ALMEIDA ENGLER, J., TIRÉ, C., ENGLER, G., VAN MONTAGU, M. and GHEYSEN, G., 1993, Induction patterns of an extensin gene in tobacco upon nematode infection, *Plant Cell*, **5**, 1697–1710.

NIEBEL, A., HEUNGENS, K., BARTHELS, N., INZE, D., VAN MONTAGU, M. and GHEYSEN, G., 1995, Characterization of a pathogen induced potato catalase and its systemic expression upon nematode and bacterial infection, *Molecular Plant–Microbe Interactions*, **8**, 371–378.

NIEBEL, A., DE ALMEIDA ENGLER, J., HEMERLY, A., FERREIRA, P., VAN MONTAGU, M. and GHEYSEN, G., 1996, Induction of cdc2a and cyc1At expression in *Arabidopsis thaliana* during early phases of nematode-induced feeding cell formation, *The Plant Journal*, **10**, 1037–1043.

OPPERMAN, C.H., TAYLOR, C.G. and CONKLING, M.A., 1994, Root-knot nematode-directed expression of a plant root-specific gene, *Science*, **263**, 221–223.

OWEN, M., GANDECHA, A., COCKBURN, B. and WHITELAM, G., 1992, Synthesis of a functional anti-phytochrome single-chain Fv protein in transgenic tobacco, *Biotechnology*, **10**, 790–794.

PHILLIPS, J., ARTSAENKO, O., FIEDLER, U., HORSTMANN, C., MOCK, H.P. and CONRAD, U., 1997, Seed-specific immunomodulation of abscisic acid activity induces a developmental switch, *The EMBO Journal*, **16**, 4489–4496.

RAY, C., ABBOTT, A.G. and HUSSEY, R.S., 1994, Trans-splicing of a *Meloidogyne incognita* mRNA encoding a putative esophageal gland protein, *Molecular and Biochemical Parasitology*, **68**, 83–101.

RIABOWOL, K.T., VOSATKA, R.J., ZIFF, E.B., LAMB, N.J. and FERAMISCO, J.R., 1988, Microinjection of fos-specific antibodies blocks DNA synthesis in fibroblast cells, *Molecular and Cellular Biology*, **8**, 1670–1676.

ROSSO, M.N., SCHOUTEN, A., ROOSIEN, J., BORST-VRENSSEN, A.W.M., HUSSEY, R.S., GOMMERS, F.J., *et al.*, 1996, Expression and functional characterization of a single chain Fv antibody directed against secretions involved in the plant nematode infection process, *Biochemical and Biophysical Research Communications*, **220**, 255–263.

RUNGGER, D., RUNGGER-BRANDLE, E., CHAPONNIER, C. and GABBIANI, G., 1979, Intranuclear injection of anti-actin antibodies into *Xenopus* oocytes blocks chromosome condensation, *Nature*, **282**, 320–321.

SCHOTS, A., DE BOER, J., SCHOUTEN, A., ROOSIEN, J., ZILVERENTANT, J.F., POMP, H., *et al.*, 1992, 'Plantibodies': a flexible approach to design disease resistance, *Netherlands Journal of Plant Pathology*, **98** supplement 2, 183–191.

SCHOUTEN, A., ROOSIEN, J., VAN ENGELEN, F.A., DE JONG, G.A.M., BORST-VRENSSEN, A.W.M., ZILVERENTANT, J.F., *et al.*, 1996, The C-terminal KDEL sequence increases the expression level of a single-chain antibody designed to be targeted to both the cytosol and the secretory pathway in transgenic tobacco, *Plant Molecular Biology*, **30**, 781–793.

SIJMONS, P.C., ATKINSON, H.J. and WYSS, U., 1994, Parasitic strategies of root nematodes and associated host cell responses, *Annual Review Phytopathology*, **32**, 235–259.

SMANT, G., GOVERSE, A., STOKKERMANS, J.P.W.G., DE-BOER, J.M., POMP, H., ZILVERENTANT, J.F., *et al.*, 1997, Potato root diffusate-induced secretion of soluble, basic proteins originating from the subventral esophageal glands of potato cyst nematodes, *Phytopathology*, **87**, 839–845.

SMANT, G., STOKKERMANS, J.P.W.G., YAN, Y., DE BOER, J.M., BAUM, T.J., WANG, X., *et al.*, 1998, Endogenous cellulases in animals: isolation of β-1,4-endoglucanase genes from two species of plant-parasitic cyst nematodes, *Proceedings of the National Academy of Sciences USA*, **95**, 4906–4911.

TAVLADORAKI, P., BENVENUTO, E., TRINCA, S., DE MARTINIS, D., CATTANEO, A. and GALEFFI, P., 1993, Transgenic plants expressing a functional single-chain Fv antibody are specifically protected from virus attack, *Nature*, **366**, 469–472.

VAN DER EYCKEN, W., DE ALMEIDA ENGLER, J., INZE, D., VAN MONTAGU, M. and GHEYSEN, G., 1996, A molecular study of root-knot nematode induced feeding sites, *The Plant Journal*, **9**, 45–54.

VAN ENGELEN, F.A., SCHOUTEN, A., MOLTHOFF, J.W., ROOSIEN, J., SALINAS, J., DIRKSE, W.G., *et al.*, 1994, Coordinate expression of antibody subunit genes yields high levels of functional antibodies in roots of transgenic tobacco, *Plant Molecular Biology*, **26**, 1701–1710.

VOSS, A., NIERSBACH, M., HAIN, R., HIRSCH, H.J., LIAO, Y.C., KREUZALER, F. and FISCHER, R., 1995, Reduced virus infectivity in *N. tabacum* secreting a TMV-specific full size antibody, *Molecular Breeding*, **1**, 15–26.

WYSS, U. 1981, Ectoparasitic root nematodes: feeding behavior and plant cell responses, in ZUCKERMAN, B.M. and ROHDE, R.A. (Eds), *Plant Parasitic Nematodes*, pp. 255–279, New York: Academic Press.

WYSS, U., 1992, Observations on the feeding behaviour of *Heterodera schachtii* throughout development, including events during moulting, *Fundamental and Applied Nematolology*, **15**, 75–89.

12

Antibodies in Plant Biotechnology –Summary and Outlook

ANGELIKA ZIEGLER AND KAREN HARPER

12.1 Introduction

Man has been genetically improving crops for thousands of years. Cultivated plants such as bread wheat, oilseed rape and potato are the result of continuous breeding efforts over a long time and no longer resemble their original parents (Forster *et al.*, 1997; Käppeli and Auberson, 1998). Methods used include simple selection and deliberate matings in combination with selection. This has resulted in plant species with better yields, faster growth and less susceptibility to disease.

Most of our basic understanding of genetics is derived from crop breeding and plant genetics (Timberlake, 1998). Mendel (1866) first described the laws of inheritance of parental traits in experiments with the garden pea. The pace of development has increased with the use of molecular tools, such as molecular markers and advances in transformation technology (Taylor, 1998). The first report of a plant transformation was in 1984 by Horsch *et al.* Since then, plant biotechnology has been developing quickly, and heterologous DNA can now be reliably incorporated into the plant genome. This has made agriculture more attractive for the large-scale production of biomolecules. Using plant transformation technology, single genes of interest can be transferred to crop plants with little disruption to the plant genome.

It is predicted that the development over the next few years will be even more rapid because of our better understanding of biochemical pathways and the effects of altering components within a pathway. In addition, there is a greater understanding of the biochemical processes involved in transgene insertion, and the efficiency of transgenic plant production has been improved.

The application of biotechnology can produce plants with new traits that would lead to improvements in food and animal feed and the plant's ability to defend itself against pathogens, pests or environmental stress. Also, products such as plastics and fuels may be synthesized in plants. Other possibilities include the production of pharmaceuticals. For example, edible vaccines and antibodies for use in therapy and disease prevention in humans and animals may be manufactured in plants more cheaply than at present. Transgenic plants may also be useful tools in bioremediation.

Regulatory genes are becoming valuable tools for metabolic engineering. For example, the activity of the transgene may be switched on or off, or the transgene may be expressed in specific organs of the plant, or at a specific time during development of the plant. Increasingly, more control is possible over the spatial or temporal control of gene expression, and multi-step metabolic pathways can be assembled in plants (Padgette *et al.*, 1997).

This chapter gives an overview of applications of antibodies in plant biotechnology and discusses some possible future uses.

12.2 Expression of foreign genes in plants

There are two different ways of expressing foreign genes in plants:

1. By stable integration of a transgene of interest into the plant genome. This can be done using the bacterium *Agrobacterium tumefaciens* (Bevan, 1984; Zambryski, 1988; effective with dicotyledonous plants) or Biolistics (using high-velocity microprojectiles; Klein *et al.*, 1987).
2. By using plants as hosts for the transient expression of foreign genes using plant pathogenic viruses as vectors for gene delivery (Chapman *et al.*, 1992; Scholthof *et al.*, 1996; Lacomme *et al.*, 1998; Porta and Lomonossoff, 1998).

The production of stably transformed plants is the more labour-intensive method, but when a stable strain has been produced, often large amounts of the product can be obtained. Transient expression systems have been shown to be useful, for example, as peptide or epitope delivery systems (Johnson *et al.*, 1997; Lacomme *et al.*, 1998; Porta and Lomonossoff, 1998). The transgene is expressed only when the plant is infected by the engineered virus. The system is rapid and convenient. High levels of expression of the foreign gene can be achieved, and large numbers of different constructs can be analysed easily in a short time.

One disadvantage of the transient system is the potential instability of the foreign gene and deletions are common. Also, care has to be taken to ensure that the engineered virus does not present a threat to the environment.

Transgenic plants can be utilized in the following ways:

1. To engineer resistance of the plant to pathogens (for example, plant viruses; Nelson *et al.*, 1988), pests (for example, insects; for a review see Jouanin *et al.*, 1998) or environmental stress (expression of polyols for osmoprotection; Sheveleva *et al.*, 1997).
2. To effect the production of a plant enzyme, hormone or other molecule normally made in the plant in order to manipulate endogenous biochemical pathways; for example, to produce speciality oils (Budziszewski *et al.*, 1996), starches of different composition (Herbers and Sonnewald, 1996) or improved food and animal feed (for example, by correcting amino acid deficiencies; Krebbers *et al.*, 1993).
3. To produce novel molecules not normally made by the plant. These include, for example, pharmaceuticals (Arntzen, 1997), antibodies for applications in humans or animals (for example, Ma and Hein, 1995) or plastics (Nawrath *et al.*, 1994, 1995). Such transgenic plants can be grown on a large scale to make bulk-quantities of desired biomolecules. For example, the production of modified oils in transgenic plants has

been particularly successful with the first transgenic crop with a modified seed composition (a lauric-oil rapeseed) cultivated commercially in 1995 (for a review, see Murphy, 1996).

12.3 Expression of antibodies in plants

Many of the aims mentioned above can be realized through the use of recombinant antibodies. Hiatt *et al.* (1989) were the first to show that whole antibodies can be assembled in transgenic plants and retain function. Work followed showing the expression of Fab fragments (De Neve *et al.*, 1993) and single-chain Fv (scFv) in transgenic plants (see Chapter 7).

For passive immunotherapy (for example, oral and gastrointestinal applications), bulk quantities of antibodies or antibody fragments are needed, and plants are well suited to provide these quantities (Ma *et al.*, 1998). Transgenic plants therefore offer a good alternative to the traditionally used microbial fermentation and animal cell culture approaches. Chapter 7 discusses the antibody yields obtained from plants. Protein farming using transgenic plants allows the production of biomolecules without the risk of contamination by mammalian viruses or endotoxins. Plants themselves do not possess an immune system, and plant viruses do not pose a danger to humans or animals. Therefore, the purification process for plant-produced proteins, including antibodies, should be less complex.

When antibodies are synthesized in mammals, they are usually glycosylated. The carbohydrates attached to antibodies synthesized in plants differ from those in mammals (for details, see Chapter 7). If the antibody is to be used in human therapy, the presence of plant-specific carbohydrates might increase the immunogenicity of the recombinant antibody (Ma and Hein, 1995). Therefore, glycosylation needs to be carefully considered (Wright and Morrison, 1997). Different requirements apply, depending on the projected use of the plant-produced antibody. Antibodies produced in field crops are already entering clinical trials and account for over one-third of products in clinical trials approved by the American Food and Drug Administration (Gibbs, 1997). Some examples are shown in Table 12.1.

Table 12.1 Examples of plant-produced therapeutic antibodies

Company	Product
Agracetus (Middleton, Wisconsin, USA)	Anti-HSV2 antibody in soybean
Planet Biotechnology, Mountain View, California, USA	Secretory IgA plantibody that blocks adherence of *Streptococcus mutans* to teeth (CaroRX)
Biosource Technologies, Inc., Vacaville, California, USA	Use of virus vector to produce, for example, antibody against non-Hodgkin's lymphoma
Guy's Hospital, London, UK	IgG and SIgA produced in tobacco plants for passive immunotherapy

Sources: Gibbs (1997); Dixon and Arntzen (1997).

12.3.1 Regulatory elements and stability of antibodies

The genetic stability of each transgenic expression system should be monitored and optimized. It is also desirable to control gene expression spatially and temporally. New tissue-specific, developmental and environmental enhancer and promoter elements are being studied for this reason (Gallie, 1998). The effects of cellular targeting on the expression levels of antibodies in plants are far from resolved. Although it was shown that signals such as the endoplasmic reticulum (ER) retention signal can have a positive effect on the level of expression of some antibodies, this does not apply to all antibodies. The cellular location and the amount of antibody that accumulates, as well as the antibody functionality in the cytoplasm, are very much dependent on the intrinsic properties of each antibody. Cattaneo and Visintin (1998) have developed selection schemes to assess scFv *in vivo* for their intracellular expression and function using a yeast two-hybrid system. However, there are still many questions left that need to be addressed, and there are as yet no general rules that can be applied to antibody expression in plants.

12.3.2 Antibody harvest and downstream processing

If the plant-produced antibodies are intended for *ex situ* applications, they have to be purified from the plant tissues. Plants chosen for antibody production for human or animal applications should not produce toxic proteins. Therefore, the choice of the plant species and plant organ in which the antibody is to be expressed is very important.

The purification of recombinant antibodies from plants can be done by standard biochemical techniques. Small peptide tags (for example, a His tag; see Chapter 4) could be added genetically to the antibodies to facilitate purification. Possible impurities can include biologically active plant metabolites or alkaloids, plant macromolecules such as DNA, polysaccharides, lipids, secondary metabolites or proteins. These can have a toxic effect (for example, nicotine from tobacco) or can cause sensitization of humans to the plant antigens. Pesticides and herbicides may also be present, if plants have been treated in the field. Furthermore, microbiological contamination may occur during the purification process (for example, by *Aspergillus flavus* or *Clostridium botulinum*; Miele, 1997).

Antibodies secreted to the apoplast should be easier to purify, they can make up 10 to 13 per cent of the intercellular fluid (ICF), and fewer plant-derived components are found in the ICF (Voss *et al.*, 1995).

The purification protocol depends on the intended use for the antibody, and the final product should be comparable with the counterpart produced by traditional methods. The challenge is to develop purification protocols that are quick, cost-effective and efficient in producing the final product with as few as possible contaminants, and without compromising antibody activity and yield. Because of the biological nature of the system, variability in yield and purity is possible from batch to batch and field to field due, for example, to the growth conditions. Product consistency has to be ensured, so mutations in the transgene sequence must be identified early. Monitoring must be implemented at every stage of the production process.

12.4 Future developments

For further development, more knowledge of metabolic regulation in plants will help to decide how to target enzymes at appropriate points within biochemical pathways and to achieve the desired effect with as few side effects as possible. With the help of biotechnology and transgenic technology, it will be possible to move from empirical methods to the design of new traits, in a predictable way.

In addition to the availability of promoters and transfer methods, and the ability to control regulatory master switches, target antigens of interest and highly specific antibodies that recognize these antigens are required.

Antibody phage display technology allows the isolation of new antibodies of predefined specificity and access to the genes that express them (Winter *et al.*, 1994). Existing hybridoma lines are also a valuable source of antibody genes (see Chapter 2).

In Chapter 10, an example of antibody-mediated virus resistance is described. Plantibodies may offer an alternative to pathogen derived resistance where there is concern about the release of virus genes into the environment. For example, coat-protein mediated resistance (Nelson *et al.*, 1988) can result in heteroencapsidation or recombination when the transgenic plant is infected by a plant virus (Tepfer, 1993). In order to create plants with antibody-mediated virus resistance, antibodies that recognize viral coat protein have been used (for example, Tavladoraki *et al.*, 1993). Most plant viruses are RNA viruses that multiply in the plant cell cytoplasm. To interfere with virus multiplication, antibodies have to be functional in the cytoplasm. This requirement is not easily achieved (see Chapter 10). Advances in functional cytoplasmic antibody expression are needed to make the concept of antibody-mediated virus resistance more useful. Antibodies to non-structural viral proteins (see Chapter 10) and proteins involved in virus movement and transmission are the next sources of resistance to be exploited. To engineer resistance to agents other than viruses (for example, bacteria, phytoplasmas) possible targets for antibody-derived disease resistance include cell-wall modifying enzymes, phytotoxins or signalling molecules of pathogens (Whitelam and Cockburn, 1996; Griep *et al.*, 1998; see also Chapter 11).

A reduction in use of pesticides would make agriculture more environmentally friendly and sustainable. For this it is necessary to improve the resistance of crops to pests. So far, engineered insect resistance is achieved using *Bacillus thuringiensis* toxin-encoding genes or plant-derived genes such as proteinase inhibitors or lectins (Jouanin *et al.*, 1998). Unfortunately, insects have a high capacity for developing resistance, even to *B. thuringiensis*-based insecticides (Shelton *et al.*, 1993). Therefore, antibodies may be a valuable alternative. However, creating pest and insect resistant plants with the help of antibodies requires more insight into the mechanisms of the interactions of the pests with the plant, at a molecular level. In Chapter 11 work is described that aims to create the basics for the engineering of resistance to nematodes using antibodies.

To allow the modulation of plant cell antigens by their interaction with a specific antibody in a suitable plant cell compartment, appropriate target genes and antibodies have to be chosen. Again, antibody phage display technology will be a helpful tool for selecting useful antibodies. The feasibility of this approach was shown by selecting scFv against plant enzymes from phage display libraries. De Jaeger *et al.* (1997) have selected scFv against dihydroflavonol 4-reductase, a key enzyme in the flavonoid biosynthetic pathway, and Ziegler *et al.* (1997) have selected scFv against enoyl acyl carrier protein reductase from oilseed rape, an enzyme that catalyses an essential step in the *de novo* fatty acid biosynthesis.

It is important to characterize the role of the targeted enzymes in order to allow a directed and efficient manipulation of their activity. Important goals include the modification of plant lipid synthesis (for example, in oilseed rape, soybean and sunflower). Metabolic partitioning in transgenic plants to favour accumulation of a desirable product can be done by removing certain products from the reaction equilibrium (for example, by sequestering the target away from its site of action) or by manipulating the levels of signal metabolites and/or phytohormones (Herbers and Sonnewald, 1996). Antibodies are well suited for these applications. Chapter 9 shows in detail how the antibody approach can be used to get more insight into hormone metabolism and function in plants. The physiological effects of hormones can be studied, helping our understanding of the regulation of biochemical pathways (Chapter 9; see also Owen *et al.*, 1992). The immunomodulation approach is a useful tool to explore other aspects of plant biochemistry.

Removal of pollutants from soil and water is another area where the potential of transgenic plants has been recognized in recent years, and phytoremediation is increasingly seen as a cost-effective alternative (Raskin, 1996; Boyajian and Carreira, 1997; Salt *et al.*, 1998). Plants are capable of absorbing heavy metals and a variety of toxic and potentially carcinogenic organic compounds (Goel *et al.*, 1997). Antibody fragments against organic pollutants (including the herbicides atrazine and paraquat) have been cloned (Byrne *et al.*, 1996), and expressed in transgenic plants (Longstaff *et al.*, 1998). The plant-produced antibodies were shown to retain specificity and affinity of the parent monoclonal antibodies. The antibody-producing plants are proposed to decrease contaminants in soil. Alternatively, the plant-produced antibodies can be purified and immobilized on solid supports, to remove the pollutants from water. For use in such industrial applications, the antibody stability has to be improved to withstand non-physiological conditions (Dooley *et al.*, 1998).

'Biofarming' is the crop plant based production of industrial or therapeutic biomolecules. Antibody concentrations of around 1 per cent of total soluble protein (TSP) are sufficient to make their expression in plants cost-effective. Chapter 8 discusses the use of plants as biofactories for the production of antibodies. Different plant species (and also different plant organs) are studied for their usefulness in the production and long-term storage of antibodies (for example, tobacco, potato). Recombinant antibodies seem stable when expressed in plant organs such as seeds or potato tubers (Artsaenko *et al.*, 1998). Antibodies produced in potato tubers were still active after 18 months of tuber storage. Antibodies (including IgG, secretory IgA (SIgA), scFv) can be produced in plants in large quantities, and already therapeutic antibodies produced in plants are undergoing human trials (see Table 12.1). SIgA are more stable than other immunoglobulins because they are polymers consisting of between two to five monomers, and also contain a secretory component (Austyn and Wood, 1993). SIgA are the predominant antibodies of the mucosal immune system and they are resistant to proteolytic digestion. Therefore, they are good candidates for oral and gastrointestinal applications (for example, passive immunization). Plant SIgA have been shown to protect against oral streptococcal colonization for at least four months (Ma *et al.*, 1998). Requirements for the purification of these antibodies are not as strict as for antibodies destined, for example, for intravenous use. Antibodies present in edible plants can be used with little purification. Food-formulated SIgA can be produced to combat tooth decay (Ma and Hein, 1995; Ma *et al.*, 1995, 1998).

scFv are also being produced in plants for use in human therapy. They are especially useful in applications in which the antibody constant region is not needed, and they can penetrate their target tissues (for example, solid tumours) better because of their small size. scFv are also less immunogenic in humans than whole antibodies.

Finally, plant-produced antibodies are by no means limited to use in human or animal therapy. Plant-produced antibodies can also be used in diagnosis.

References

ARNTZEN, C.J., 1997, High-tech herbal medicine: plant-based vaccines, *Nature Biotechnology*, **15**, 221–222.

ARTSAENKO, O., KETTIG, B., FIEDLER, U., CONRAD, U. and DÜRING, K., 1998, Potato tubers as a biofactory for recombinant antibodies, *Molecular Breeding*, **4**, 313–319.

AUSTYN, J.M. and WOOD, K.J., *Principles of Cellular and Molecular Immunology*, Oxford University Press, 1993.

BEVAN, M., 1984, Binary Agrobacterium vectors for plant transformation, *Nucleic Acids Research*, **12**, 8711–8721.

BOYAJIAN, G.E. and CARREIRA, L.H., 1997, Phytoremediation: a clean transition from laboratory to marketplace?, *Nature Biotechnology*, **15**, 127–128.

BUDZISZEWSKI, G.J., CROFT, K.P.C. and HILDEBRAND, D.F., 1996, Uses of biotechnology in modifying plant lipids, *Lipids*, **31**, 557–569.

BYRNE, F.R., GRANT, S.D., PORTER, A.J. and HARRIS, W.J., 1996, Cloning, expression and characterization of a single-chain antibody specific for the herbicide atrazine, *Food and Agricultural Immunology*, **8**, 19–29.

CATTANEO, A. and VISINTIN, M., 1998, Phage display technology: antibodies, proteins and peptides, International Symposium at the School of Molecular and Medical Biosciences, Cardiff University, June 1998.

CHAPMAN, S.N., KAVANAGH, T.A. and BAULCOMBE, D.C., 1992, Potato virus x as a vector for gene expression in plants, *Plant J.*, **2**, 549–557.

DE JAEGER, G., BUYS, E., EECKHOUT, D., BRUYNS, A.-M., DE NEVE, M., DE WILDE, C., *et al.*, 1997, Use of phage display for isolation and characterization of single-chain variable fragments against dihydroflavonol 4-reductase from *Petunia hybrida*, *FEBS Letters*, **403**, 116–122.

DE NEVE, M., DE LOOSE, M., JACOBS, A., VAN HOUDT, H., KALUZA, B., WEIDLE, U., *et al.*, 1993, Assembly of an antibody and its derived antibody fragment in *Nicotiana* and *Arabidopsis*, *Transgenic Research*, **2**, 227–232.

DIXON, R.A. and ARNTZEN, C.J., 1997, Transgenic plant technology is entering the era of metabolic engineering, *Trends in Biotechnology*, **15**, 441–444.

DOOLEY, H., GRANT, S.D., HARRIS, W.J. and PORTER, A.J., 1998, Stabilization of antibody fragments in adverse environments, *Biotechnology and Applied Biochemistry*, **28**, 77–83.

FORSTER, B.P., LEE, M.A., LUNDQVIST, U., MILLAM, S., VAMLING, K. and WILSON, T.M.A., 1997, Genetic engineering of crop plants: from genome to gene, *Experimental Agriculture*, **33**, 15–33.

GALLIE, D.R., 1998, Controlling gene expression in transgenics, *Current Opinion in Plant Biology*, **1**, 166–172.

GIBBS, W.W., 1997, Biotechnology – Plantibodies, *Scientific American*, **277**, 44.

GOEL, A., KUMAR, G., PAYNE, G.F. and DUBE, S.K., 1997, Plant cell biodegradation of a xenobiotic nitrate ester, nitroglycerin, *Nature Biotechnology*, **15**, 174–177.

GRIEP, R.A., VAN TWISK, C., VAN BECKHOVEN, J.R.C., VAN DER WOLF, J.M. and SCHOTS, A., 1998, Development of specific recombinant monoclonal antibodies against the lipopolysaccharide of Ralstonia solanacearum race 3, *Phytopathology*, **88**, 795–803.

HERBERS, K. and SONNEWALD, U., 1996, Manipulating metabolic partitioning in transgenic plants, *Trends in Biotechnology*, **14**, 198–205.

HIATT, A., CAFFERKEY, R., and BOWDISH, K., 1989, Production of antibodies in transgenic plants, *Nature*, **342**, 76–78.

HORSCH, R., FRALEY, R., ROGERS, S., SANDERS, P., LLOYD, A. and HOFFMAN, N., 1984, Inheritance of functional foreign genes in plants, *Science*, **223**, 496–498.

JOHNSON, J., LIN, T., LOMONOSSOFF, G., 1997, Presentation of heterologous peptides on plant viruses: genetics, structure and function, *Annual Review of Phytopathology*, **35**, 67–86.

JOUANIN, L., BONADE-BOTTINO, M., GIRARD, C., MORROT, G. and GIBAND, M., 1998, Transgenic plants for insect resistance, *Plant Science*, **131**, 1–11.

KÄPPELI, O. and AUBERSON, L., 1998, How safe is safe enough in plant genetic engineering?, *Trends in Plant Science*, **3**, 276–281.

KLEIN, T.M., WOLF, E.D., WU, R. and SANFORD, J.C., 1987, High-velocity microprojectiles for delivering nucleic acids into living cells, *Nature*, **327**, 70–73.

KREBBERS, E., VAN ROMPAEY, J. and VANDERKERCKHOVE J., 1993, Expression of modified seed storage proteins in transgenic plants, in *Transgenic Plants: Fundamentals and Applications*, ed. Hiatt, A., Marcel Dekker Inc., New York, USA, pp. 37–60.

LACOMME, C., SMOLENSKA, L. and WILSON, T.M.A., 1998, Genetic engineering and the expression of foreign peptides or proteins using plant virus-based vectors, in *Genetic Engineering: Principles and Methods, Vol. 20*, ed. Setlow, J.K., Plenum Press, New York, pp. 225–237.

LONGSTAFF, M., NEWEL, C.A., BOONSTRA, B., STRACHAN, G., LEARMONTH, D., HARRIS, W.J., *et al.*, 1998, Expression and characterization of single-chain antibody fragments produced in transgenic plants against the organic herbicides atrazine and paraquat, *Biochimica et Biophysica Acta*, **1381**, 147–160.

MA, J.K.C. and HEIN, M.B., 1995, Immunotherapeutic potential of antibodies produced in plants, *Trends in Biotechnology*, **13**, 522–527.

MA, J.K.C., HIATT, A., HEIN, M., VINE, N.D., WANG, F., STABILA, P., *et al.*, 1995, Generation and assembly of secretory antibodies in plants, *Science*, **268**, 716–719.

MA, J.K.C., HIKMAT, B.Y., WYCOFF, K., VINE, N.D., CHARGELEGUE, D., YU, L., *et al.*, 1998, Characterization of a recombinant plant monoclonal secretory antibody and preventive immunotherapy in humans, *Nature Medicine*, **4**, 601–606.

MENDEL, G., 1866, Experiments in plant hybridization (English translation of: Versuche über Pflanzenhybriden, 1958), in *Principles of Genetics*, eds E.W. Sinnot, L.C. Duncan and Th. Dobzhansky, McGraw-Hill, New York.

MIELE, L., 1997, Plants as bioreactors for biopharmaceuticals: regulatory considerations, *Trends in Biotechnology*, **15**, 45–50.

MURPHY, D.J., 1996, Engineering oil production in rapeseed and other oil crops, *Trends in Biotechnology*, **14**, 206–213.

NAWRATH, C., POIRIER, Y. and SOMERVILLE, C., 1994, Targeting of the polyhydroxybutyrate biosynthetic pathway to the plastids of *Arabidopsis thaliana* results in high levels of polymer accumulation, *Proceedings of the National Academy of Sciences USA*, **91**, 12760–12764.

NAWRATH, C., POIRIER, Y. and SOMERVILLE, C., 1995, Production of polyhydroxyalkanoates in plants, in *Program and Abstracts for the International Symposium on Engineering Plants for Commercial Products and Applications, Session III: Industrial Uses*, Lexington, Kentucky, USA.

NELSON, R.S., MCCORMICK, S.M. and DELANNAY, X., 1988, Virus tolerance, plant growth and field performance of transgenic tomato plants expressing coat protein from tobacco mosaic virus, *Bio/Technology*, **6**, 403–409.

OWEN, M., GANDECHA, A., COCKBURN, B. and WHITELAM, G., 1992, Synthesis of a functional anti-phytochrome single-chain Fv protein in transgenic tobacco, *Bio/Technology*, **10**, 790–794.

PADGETTE, S.R., GRUYS, K.J., MITSKY, T.A., TRAN, M., TAYLOR, N.B., SLATER, S.C. and KISHORE, G.M, 1997, Strategies for production of polyhydroxyalkanote polymers in plants, *Plant Physiology*, **114**, 297.

PORTA, C. and LOMONOSSOFF, G.P., 1998, Scope for using plant viruses to present epitopes from animal pathogens, *Reviews in Medical Virology*, **8**, 25–41.

RASKIN, I., 1996, Plant genetic engineering may help with environmental cleanup, *Proceedings of the National Academy of Sciences USA*, **93**, 3164–3166.

SALT, D.E., SMITH, R.D. and RASKIN, I., 1998, Phytoremediation, *Annual Review of Plant Physiology and Plant Molecular Biology*, **49**, 643–668.

SCHOLTHOF, H.B., SCHOLTHOF, K.B.G. and JACKSON, A.O., 1996, Plant-virus gene vectors for transient expression of foreign proteins in plants, *Annual Review of Phytopathology*, **34**, 299–323.

SHELTON, A.M., ROBERTSON, J.L., TANG, J.D., PEREZ, C., EIGENBRODE, S.D., PREISLER, H.K., *et al.*, 1993, Resistance of the diamondback moth (Lepidoptera, Plutellidae) to *Bacillus thuringiensis* subspecies in the field, *Journal of Economic Entomology*, **86**, 697–705.

SHEVELEVA, E.V., BOHNERT, H.J. and JENSEN, R.G., 1997, Metabolic engineering of polyol production in *Nicotiana tabacum*, *Plant Physiology*, **114**, 298.

SMITH, M.D., 1996, Antibody production in plants, *Biotechnology Advances*, **14**, 267–281.

TAVLADORAKI, P., BENVENUTO, E., TRINCA, S., MARTINIS, D.D., CATTANEO, A. and GALEFFI, P., 1993, Transgenic plants expressing a functional single-chain Fv antibody are specifically protected from virus attack, *Nature*, **366**, 469–472.

TAYLOR, C.B., 1998, Factories of the future? Metabolic engineering in plant cells, *The Plant Cell*, **10**, 641–644.

TEPFER, M., 1993, Viral genes and transgenic plants, *Bio/Technology*, **11**, 1125–1132.

TIMBERLAKE, W., 1998, Agricultural genomics comes of age, *Nature Biotechnology*, **16**, 116–117.

VOSS, A., NIERSBACH, M., HAIN, R., HIRSCH, H.J., LIAO, Y.C., KREUZALER, F. and FISCHER, R., 1995, Reduced virus infectivity in *Nicotiana tabacum* secreting a TMV-specific full-size antibody, *Molecular Breeding*, **1**, 39–50.

WHITELAM, G.C. and COCKBURN, W., 1996, Antibody expression in transgenic plants, *Trends in Plant Science*, **1**, 268–272.

WINTER, G., GRIFFITHS, A.D., HAWKINS, R.E. and HOOGENBOOM, H.R., 1994, Making antibodies by phage display technology, *Annual Review of Immunology*, **12**, 433–455.

WRIGHT, A. and MORRISON, S.L., 1997, Effect of glycosylation on antibody function: implications for genetic engineering, *Trends in Biotechnology*, **15**, 26–32.

ZAMBRYSKI, P., 1988, Basic processes underlying Agrobacterium-mediated DNA transfer to plant cells, *Annual Review of Genetics*, **22**, 1–30.

ZIEGLER, A., MACINTOSH, S.M., TORRANCE, L., SIMON, W. and SLABAS, A.R., 1997, Recombinant antibody fragments that detect enoyl acyl carrier protein reductase in *Brassica napus*, *Lipids*, **32**, 805–809.

Index